The Ghost of the Executed Engineer

LOREN R. GRAHAM

The Ghost of the Executed Engineer

Technology and the Fall of the Soviet Union

HARVARD UNIVERSITY PRESS
CAMBRIDGE, MASSACHUSETTS
LONDON, ENGLAND

First Harvard University Press paperback edition, 1996

Library of Congress Cataloging-in-Publication Data

Graham, Loren R.
  The ghost of the executed engineer : technology and the fall
of the Soviet Union / Loren R. Graham.
    p. cm. — (Russian Research Center studies ; 87)
  Includes bibliographical references and index.
  ISBN 0-674-35436-2 (cloth)
  ISBN 0-674-35437-0 (pbk.)
  1. Pal'chinskiĭ, Petr Ioakimovich, d. 1929. 2. Engineers —
Soviet Union—Biography. 3. Engineers—Soviet Union—Social
conditions. 4. Technology and state—Soviet Union.
5. Technology—Soviet Union—History. I. Title. II. Series.
TA140.P25G73 1993
609.47 ' 09 ' 04—dc20   93-1119
                                                              CIP

Russian Research Center Studies, 87

To the memory of Victor Lincoln Albjerg,
my teacher of the humanities at an American
engineering institution. He shared Palchinsky's
vision, and his influence on hundreds of students
contributes to the United States' industrial
and academic strengths.

Contents

Illustrations

Blast furnace at Magnitogorsk
Courtesy of estate of Margaret Bourke-White

Map of the White Sea Canal
Belomor: An Account of the Construction of the New Canal between the White Sea and the Baltic Sea (New York: Harrison Smith and Robert Hass, 1935)

Prison workers at the White Sea Canal
M. Gor'kii, L. Averbakh, S. Firin, Belomorsko-Baltiiskii Kanal imeni Stalina (Moscow: OGIZ, 1934)

N. I. Khrustalev, chief engineer of the White Sea Canal
M. Gor'kii, L. Averbakh, S. Firin, Belomorsko-Baltiiskii Kanal imeni Stalina (Moscow: OGIZ, 1934)

First steamers on the White Sea Canal
M. Gor'kii, L. Averbakh, S. Firin, Belomorsko-Baltiiskii Kanal imeni Stalina (Moscow: OGIZ, 1934)

Construction of the Baikal-Amur Railway
TASS/Sovfoto

Aerial view of the Chernobyl nuclear power plant
TASS/Sovfoto

Monument at the end of the White Sea Canal
Intourist/Sovfoto

Prologue

This book attempts to help explain why the Soviet Union failed to become a modern industrialized country. It starts with the life story of a remarkable Russian engineer, Peter Palchinsky, who saw clearly at the beginning of Soviet industrialization the mistakes that were being made, and tried to rectify them. The story of Palchinsky serves as a parable for the remainder of the book, which presents an analysis of Soviet attitudes toward industry and technology during the sixty years after Palchinsky's death. Palchinsky's critique of the misuse of technology and squandering of human energy continued to haunt the Soviet Union until its demise at the end of 1991.

Interwoven with these two parts of the book is a personal story of my quest for more than thirty years to unravel the riddle of Palchinsky and his role in the efforts to industrialize the Soviet Union. Almost every textbook of Soviet history mentions the Industrial Party Trial in 1930, a prosecution of many leading Russian engineers. Few of those texts offer any information about the alleged head of the Industrial Party, Peter Palchinsky. I first learned his name while doing graduate study at Moscow University in 1960–61. My early attempts to find out more about him were frustrated by Soviet secrecy. The archives I needed to search were closed, not only to me but to all researchers, including Soviet ones. However, from the 1960s on I kept a file on Palchinsky, adding the tidbits of information that I picked up from time to time. Long before the archives became available small breakthroughs occurred, moments when I found something about Palchinsky during frequent trips to the Soviet Union, which allowed me ample opportunity to observe the failures of technology to serve the Soviet people.

One important discovery came in the early 1980s, when my colleague, Sheila Fitzpatrick, who knew of my interest in the Soviet engineers, told me there was a copy of a secret police report on the Industrial Party at the Institute of Scientific Information on the Social Sciences (INION) of the Academy of Sciences of the USSR. The difficulties I experienced in gaining access to this material exemplify the obstacles to research under the Soviet regime.

Finding such a report in the public collections of a Soviet library was highly unlikely to begin with. Usually sensitive materials were kept in Special Collections (*spetskhran*) in Soviet libraries, and were not even listed in the public catalogs. And the Soviet definition of "sensitive materials" was extremely restrictive. My own books, for example, were not listed in the catalog of the Lenin Library—the largest library in the Soviet Union—even though they were academic monographs on rather narrow subjects. There was no hope of finding the works of such well-known "enemies" of the Soviet regime as Nikolai Bukharin or Leon Trotsky. (My heart did skip a beat when in the 1970s I found a reference in the Lenin Library card catalog to "L. Trotsky." This Trotsky, alas, turned out to be an automotive engineer specializing in the design of brakes.)

The INION library is open only to researchers connected with the Academy of Sciences. As a participant in an official exchange between the Soviet and American academies I was eligible for a pass. The library is different from all others in the Soviet Union in which I have worked: cleaner, lighter, and with a freer atmosphere. To my surprise, I found two of my books in the catalog, as well as others by Western scholars working in Russian and Soviet studies. Even several of Bukharin's and Trotsky's works were listed. And the collection of materials on the 1920s was far richer than what is openly catalogued in the Lenin Library.

The relative unorthodoxy of the INION collections has an interesting history. The heart of the collection is the library of the Communist Academy of the twenties, an association of Marxist scholars that flourished before Stalinist controls clamped down on intellectual life in the Soviet Union. Their articles in the Communist Academy's journal often expressed views that were later condemned. A collection based on their work would thus be more variegated politically than the typical Soviet library.

Searching through the card catalog under the subject heading of the Industrial Party I soon found a reference to a confidential report on the early engineers by the OGPU (predecessor of the KGB). The report had been prepared for the members of the Central Committee of the Communist Party on the occasion of the Sixteenth Congress of the Party (June 26–July 13, 1930), a few months before the Industrial Party Trial. A quick glance at the report confirmed that the materials were sensitive. Though I wanted a copy of the entire manuscript, I feared that my request would be refused and that the original report would be taken away from me. I therefore made extensive notes on the report before taking it to the photoduplication department of INION, where the young woman in charge, whom I will call Nina Smirnova, knew me. To my delight, she accepted the duplication order without looking at the title or asking any questions. About a week later, I picked up my microfilm and immediately sent it back to the United States via the American Embassy. Then I returned to my note-taking from the original, secure in the knowledge that I would not lose access to this valuable source.

My apprehensions were confirmed when Nina Smirnova sought me out in the library a day or two later and demanded that I return the report. I gave the original to her but told her the copy had already been mailed home. She became very agitated, and said that the Communist Party organization at INION had become aware of my research and had forbidden me to have access to unpublished materials. She asked that I not tell anyone that I had mailed a copy of the secret police report to the United States. I replied that it seemed strange to be so concerned about an event that had occurred more than fifty years earlier. I also observed that the report had been openly listed in the catalog at INION, and that I was therefore not doing anything wrong. Ms. Smirnova replied, "It's not openly listed anymore." I expressed the hope that my research had not gotten her into trouble. She said that if I kept my mouth shut, she would be all right. We parted on good terms.

Returning to the catalog, I again looked up the reference to the report. The card listing it had disappeared, but at the bottom of the tray was a telltale piece of cardboard, showing that the card had been simply ripped out.

Soviet resistance to my efforts to learn more about Peter Palchinsky began to crumble in the late eighties. As more and more information came into my hands, I began to realize that his ideas survived his death and outlived the USSR itself. His ghost has guided me to an understanding of the failures of Soviet technology and the great cost that industrialization exacted from the Soviet people.

Grand Island, Lake Superior
June 1993

The Ghost of the Executed Engineer

The Radical Engineer 1

On a cold night in April 1928, Stalin's secret police knocked on the door of Peter Palchinsky's Leningrad apartment. When Palchinsky, a fifty-four-year-old engineer, came to the door the police announced that he was under arrest.[1] They searched his apartment and found an enormous collection of personal papers concerning his work as an engineer for more than thirty years. As the police took Palchinsky away they commanded his wife, Nina Aleksandrovna, to carry her husband's papers in bags to the police station. She heard no more about the fate of her husband for more than a year, until, on May 24, 1929, the Soviet newspaper *Izvestiia* published a short and shocking statement. Nina Aleksandrovna read that her husband had been the leader of an anti-Soviet conspiracy trying to overthrow the government and restore capitalism, that he had been convicted without trial for treason, and that immediately thereafter he was executed by a firing squad.[2]

Many years later the arrest and death of Palchinsky were briefly described by Aleksandr Solzhenitsyn in his *Gulag Archipelago*. Solzhenitsyn observed that the papers of this outstanding engineer had disappeared into the "maw" of the secret police, "once and for all, forever."[3] Until the present day little has been known of Palchinsky, although many Western historians recognized his prominent role in the industrialization and technical development of Russia during the first decades of this century. In 1982 an American historian managed to write a few paragraphs about him in an encyclopedia of Russian history, noting that "little information is available on Palchinsky, and Soviet sources are silent on him."[4]

Sixty-two years after Palchinsky's execution, on an icy Moscow day in January 1991, I was permitted to inspect a government archive that I had been denied entrance to for several decades. The Soviet Union was now in the midst of Gorbachev's reforms, and although the stocks in food in stores were almost nonexistent, *glasnost'* had brought new life to political debate and to scholarly research. Inside the archive, I found a microfilm index to the collections, but no reels for the microfilm readers. At first stymied by this technical problem, I soon noticed a neighbor jamming his finger into a reel-less microfilm roll and frantically cranking away at the reader with the other hand. I imitated him and the film became legible—just barely. After an hour or so I spotted a reference to the file of P. A. Palchinsky. Upon locating the file, I was overwhelmed by its size. Because archive rules stipulated that I could order only ten packets each day from the hundreds in the file, it quickly dawned on me that my fact-finding was turning into an odyssey. Furthermore, whenever I returned to the microfilm index I had to obey the archive rule of turning the microfilm reader off for fifteen minutes every hour so that it would cool down and not ignite the film. I soon learned that if I came to the archive early enough in the morning I could get one of the few foreign-made microfilm readers for which this rule did not apply. Over the following months and during three more research trips to Moscow, the entire collection that Nina Aleksandrovna Pal'chinskaia had hauled off to the police station slowly surfaced, like a giant fish from beneath the water.

Reading through these materials as the Soviet Union disintegrated around me, I saw that here was a clue to one of the riddles of Soviet history. Why had the USSR been unable to benefit fully from its impressive start in technological modernization? From its inception the leaders of the Soviet Union had put great emphasis on technology, launching programs of electrification, industrialization, and weapons building that inspired some Western observers and alarmed others. The Soviet efforts to exploit technology at first seemed quite successful. On Soviet soil during the Five-Year Plans launched before World War II arose the world's largest steel mills and largest hydroelectric power plants. Foreign observers and participants, from the photographer Margaret Bourke-White to the labor leader Walter Reuther, came to witness and admire the "Great Soviet Experiment."

The Soviet economy continued to lunge forward in a spasm of expansion and modernization that fascinated observers everywhere. Alexander Gerschenkron, an economic historian at Harvard University, advanced the thesis of the "advantages of backwardness," maintaining that when the Soviet Union installed factory equipment for the first time, it was the latest model, putting the USSR in a superior position to those countries that had expanded earlier and were saddled with obsolescent technology. As late as 1960, over forty years after the Russian Revolution, Robert Campbell, an economist who would become a leader among specialists on the Soviet Union, noted that the Soviet economy was growing almost twice as fast as the American one and concluded, "As long as there is a differential in the rate of growth the Russians will inevitably catch up with us, and if the differential continues at anything like its present magnitude, they will gain on us rapidly."[5]

More than two generations after the Russian Revolution, we now see that this grand effort to master technology and use it for the nation's benefit failed. Gorbachev, Yeltsin, and other recent leaders of the Soviet Union and its successor states have instead appealed for Western help in modernization. What caused this failure? The usual answer, the limitations of a centrally planned economy, is only a partial one. After all, the Soviet centrally planned economy worked well enough to build up an industrial establishment that was, in its heyday, the second largest in the world; it enabled the Soviet Union to resist and throw back Hitler's armies, and to continue to expand for many decades, both before and after World War II. It gave the Soviet Union the ability to launch the world's first artificial satellite and to put the first human being into orbit around the earth. So long as Soviet citizens had faith in their system it seemed to work fairly well, at least in comparison with other backward nations trying to modernize. Was there something about the way technology was used that contributed to the loss of faith and the consequent failure? The story of Peter Palchinsky's life and ideas about technology provides an important piece of this puzzle.[6]

The Making of a Young Engineer

Peter Akimovich Palchinsky came from a large, complicated, and troubled family. His father, Akim Fedorovich Pal'chinskii, a land surveyor and estate appraiser, married twice and had five children by his first wife, Aleksandra, and seven children by his second, Olga. Born on October 5, 1875, Peter was the oldest son, and was regarded by his siblings as the person to whom to confess difficulties and from whom to seek aid, psychological and financial. As a child Peter lived with his mother, Aleksandra, in the Volga river city of Kazan, along with his brother Fedor and his three sisters Anna, Sophia, and Elena. His four half-brothers—Ivan, Mikhail, Aleksandr, and Il'ia—and three half-sisters—Antonina, Julia, and Aleksandra—lived with his father and their mother in the city of Saratov.

Peter was an energetic youth and a bright student. After the age of eight, when his parents divorced, he saw his father rarely. He confided primarily in his mother, a member of a socially promi-nent but impecunious noble family. His mother greatly influenced Peter's early education. Under her tutelage he became a good pianist, despite his lack of natural proclivity for the instrument.[7] She also encouraged him to read in the large inherited family library. Peter spent hours buried in the novels, poetry, popular science books, and historical accounts that he found in his home. At the same time, he lamented to his mother his lack of close friends. She recognized that Peter was unusually self-contained, and she urged him to be more expressive with others.[8] But she also praised his academic accomplishments, which included mastery of French and German by the age of seventeen.[9] Later he would add English and Italian to his linguistic repertory.

In the fall of 1893 Palchinsky entered the Mining Institute in St. Petersburg, one of the elite engineering institutions of tsarist Russia. On the entrance examinations he received excellent scores: German 12 (out of 12), physics 10, mathematics 10.5, Russian 8 (he was sick the day he took this examination). Nevertheless, he ranked twentieth in his entering class of thirty-five students. He took special pride that he had been admitted to the Mining Institute without any help from influential friends or high officials.[10]

During his student years Peter lived on such a modest budget that he often did not have enough money to eat properly. His mother wrote to him: "It is a bitter experience to know that I am completely powerless in improving your situation."[11] When she became sick and died a few weeks later, Peter was reduced to living on a very small student's stipend. To supplement his income, during summer recesses he worked as a laborer on railroads, in factories, and even in coal mines in France.[12] In these occupations he developed a sympathy for the workers and for their efforts to improve pay and working conditions. Ironically, of all the members of Peter's large family, he was the only one who became a financial success, and many of the siblings' letters to Peter contain requests for money. Peter was also the one with the most robust health and steady personality, almost never yielding to the illness and despondency that ran through much of the rest of the family.

Like many young educated Russians at the turn of the century, Peter Palchinsky was attracted to radical political doctrines that promised a better society than the authoritarian and poverty-stricken one in which he was born. He soon suffered for these beliefs. Even as a student at the Mining Institute he attracted the attention of the tsarist gendarmes, who listed him as a "leader of the movement" of radical students, evidently because he was briefly the chairman of a students' assembly. This early political difficulty was a harbinger of many more in his life: he would be imprisoned five or six times and was under almost constant surveillance by the tsarist police and later by the Soviet secret police.

Peter's interest in politics and the arts was nurtured by his family. Fedor, the brother who was closest to Peter in age and who tried to help support the family, sought relief from his boring work as a low-ranking official by attending the theater, consorting with actors at receptions and parties, and volunteering to assist theater directors. On visits home to Kazan, Peter joined his brother in attending cultural events. He was also strongly attracted to the arts, but did not submerge his personality in them as Fedor seemed to do.

The other siblings, like Fedor, had difficulty finding satisfying work and adjusting to life. Peter's sister Sophia was often sick and manifested little energy. She married a Muslim citizen of Russia named Mukhamed Syzdikov, by whom she had one daughter, but the marriage ended in divorce.

The youngest sister Elena was an inveterate romantic who loved literature, music, and the theater, and dreamed of becoming an artist. She went to Brussels and Paris to attend lectures in the arts and literature. While in Paris she joined up with Russian radical exiles who influenced her politically. In 1898 Elena attended a congress of socialists in Paris, and there she discovered the writings of Karl Marx. She asked Peter to send her a copy of the Russian translation of *Capital*, but it never arrived. It is not clear whether Peter refused to buy it, or whether it was lost in the mail. Elena was forced to try to read the book in French.

Always short of funds, Elena made several trips back and forth between Brussels and Kazan, and finally returned to Kazan, where she quickly made an unhappy marriage. Fleeing that union and the tedium of provincial life, she moved to St. Petersburg, where she took a job as a bank clerk, work which she found intolerably dull.

Anna was the only one of Peter's full sisters who managed to keep her life in order. Since their mother Aleksandra was often sick, Anna, an energetic and self-controlled young woman, from an early age actually ran the home. To earn money she helped her mother run a subscription library in Kazan, based on the inherited family library that had been so important in Peter's education. After her mother's death in 1893, Anna took over the library. In 1896 the collection contained over eight thousand books in seven languages.[13] Income from the library proved insubstantial, so Anna added a bakery to her businesses. Eventually Anna married happily, and throughout her life carried on a frequent correspondence with Peter. In these letters she describes events in the lives of the other family members, and neglects only herself. The sole sign in her letters of the lack of order that seemed to plague the rest of the family is the inaccuracy of the dates; in fact, several times, in place of a date, she simply writes "Unknown date."

Among the seven half-sisters and half-brothers a pattern of difficulty often appeared. Antonina and Ivan were so frequently ill that they were unable to hold down jobs. Julia, a sweet young woman, at an early age became a teacher in a girls' school; unfortunately, her amiable nature prevented her from enforcing the necessary discipline, and her supervisors gave her low evaluations as a teacher. On a part-time basis she gave private lessons to help pay the expenses for the large family. Her greatest pleasure was playing a rented piano and singing. When she lost her job at the girls' school and was unable to continue payment of the rental for the piano, for the first and only time in her life, she appealed to Peter for money, which Peter promptly sent. Peter's correspondence from this attractive woman, who combined the romanticism of her half-sister Elena with the practicality of her half-sister Anna, ceased when she became seriously ill with tuberculosis.

Peter's half-brother Mikhail, like Peter, was interested in mining in Siberia, where he briefly worked. Also like Peter, he sympathized with the workers and became active in radical organizations. Eventually he was arrested by the tsarist police for his work in an underground workers' press, and was locked up in the Spasskii Prison in St. Petersburg. From the prison he appealed for help to Peter. Whether or not as a result of Peter's intercession, Mikhail was soon released and returned to Saratov. From there he attempted to gain admission to an engineering institute, but was rejected for "lack of evidence of trustworthiness." Soon after, he was called up for military service, and his letters to Peter ceased.

Compared to his siblings, Peter was a monument to stability and prosperity. His half-sister Julia once wrote that he was like a toy familiar to all Russian children, a wooden doll with a lead weight in its bottom that always righted itself no matter how many times it was knocked over.[14] Julia observed that Peter possessed an amazing ability to overcome difficulties without damage to his psyche. By his mid-twenties he was supporting much of the family.

On November 23, 1899, Peter married Nina Aleksandrovna Bobrishcheva-Pushkina, a member of a prominent St. Petersburg family. They lived in St. Petersburg until he graduated with honors from the Mining Institute the following year. Nina Aleksandrovna prepared herself for a move far away, since she knew that student recipients of state stipends such as her husband were required to accept government assignments.

Coal Mining in the Don Basin

The assignment Peter Palchinsky received in 1901 was to study a decline in coal production in Ukraine's Don Basin. Inadequate supplies of coal threatened the continued growth of industry after a decade of the most intensive industrial expansion in Russian history. Coal was crucial for the buildup of Russia's industrial and military might at a time of intense competition among the European powers. The Don Basin supplied almost 70 percent of Russia's coal in 1900.

As the youngest member of the investigative commission that was sent from St. Petersburg, Palchinsky was given the less prestigious task of looking into the "worker question" while the others studied mine operations. He began modestly by trying to collect information on the number of workers and mines in the Don Basin. To his amazement, the operators of the mines knew very little about their workers, not even their total numbers or the number of worker-days annually expended in each mine.[15] The mine owners also seemed uninterested in the workers' living conditions. Palchinsky decided to collect his own statistics. As he worked, he realized that such statistics were absolutely necessary to understand the workers and their productivity. At this time one of the credos of all his later activity was formed: Good industrial policy cannot be formulated in the absence of full and reliable statistics.

He worked feverishly for over two years, collecting an enormous quantity of information, including architectural drawings of workers' housing, photographs (still contained in his archive), and maps of population density and transportation networks. At the Makar'evskii Mine he found barracks that housed sixty-eight mine workers in a single room, with plank beds lined up in long rows of twenty or more and no space between them. The only way a worker could get into his bed without crawling over his fellow workers was by entering at the foot and crawling its length. At the Gorlovka Mine he found a similar situation, with forty workers to a room. At both of these mines the barracks had been constructed of brick by the mine owners, and looked rather substantial from the outside. The buildings were, indeed, superior to the mud dugouts (zemlianki) common among miners a few years earlier. Palchinsky made detailed drawings inside the barracks showing the location of each bed, the heating stove, and the sanitary facilities (usually crude outhouses). For workers with families, Palchinsky found a different housing pattern: families living four to six to a house, with one family in each room. Often these houses had dirt floors and lacked toilet facilities.[16]

No one had ever before collected this kind of information on housing conditions in the Don Basin. Palchinsky sent back his reports to St. Petersburg, with neat summary tables and drawings, and with no political comment. At first the Minister of Finance, V. I. Kovalevskii, and officials in the Ministry of Trade and Industry recognized the value of the work and even suggested that it be applied elsewhere in other branches of industry. Gradually, however, the political significance of the reports began to dawn on these officials. In 1906 Palchinsky sent a manuscript for publication in the *Mining Journal* in which he reported that even in the best housing (one-family workers' homes), out of more than 20,000 such structures 16,400 had either earthen floors or earthen roofs, or both, and were unhealthy residences, especially in the winter.[17] The manuscript caused an uproar when reported to Palchinsky's superiors and resulted in his dismissal from the investigative commission. Though sent to Siberia in what amounted to administrative exile, he was permitted to continue working as a consultant to the mining industry.

Radicalized by these experiences, Palchinsky was drawn at first to the anarchism of Peter Kropotkin (1842–1921), a Russian revolutionary of noble lineage who at the end of the nineteenth century and beginning of the twentieth wrote influential books that portrayed the possibility of a new society without exploitation or oppression.[18] Kropotkin's anarchism was a more moderate variety than that of Mikhail Bakunin (1814–1876). Instead of calling for violence, Kropotkin spoke of "mutual aid" and the benefit that would come to civilization if it would reconstruct itself on the basis of autonomous associations of agrarian and industrial producers working peacefully and cooperatively with each other. In such a system, mental and physical work would be combined, along with the advantages of countryside and city.

Just what anarchism meant to Palchinsky is difficult to discern, since he seems to have had little interest in political theory itself. He clearly disliked the exploitation that he saw in capitalist society, and he often spoke of the advantages of common ownership of the land and of cooperation among all the members of society. In his voluminous writings, however, he usually concentrated on practical topics. Scattered through those writings are many of Kropotkin's favorite phrases, such as "the satisfaction of the needs of men with the least possible waste of energy" and "the integration of labor."

The aspect of Kropotkin's teachings that seemed to appeal most to Palchinsky was his attitude toward technology. Palchinsky admitted that Kropotkin had constructed a utopia, but he was inspired by the fact that, in contrast to many utopians, Kropotkin considered technology a friend rather than an enemy.[19] Kropotkin believed that the industrial revolution of the eighteenth and nineteenth centuries was a cruel aberration in history, a temporary phase when financial capital and steam technology worked together to form an oppressive society based on centralized factories with division of labor and resulting class conflict. Yet Kropotkin believed that in the near future, new technologies, such as electricity and the telephone, would lead to novel forms of work in agriculture and industry. The advantages of small cooperatives dispersed throughout the land would become clear. The coming society was to be heterogeneous, combining a few large enterprises with a multitude of autonomous smaller ones.

Although Palchinsky would retain this commitment to a coming utopian relationship of technology and society, he did not become a loyal adherent of anarchism as a political movement. He distinguished between Kropotkin's writings on future society and the brash actions of some of his followers. Palchinsky instead concentrated his political efforts on writing articles that called for social insurance, a shorter working day, and adequate pay.[20]

In the Revolution of 1905 Palchinsky did not join the anarchists' agitations and occasional robberies, but he did give his support to the revolution, was arrested as a result, and was sentenced to exile under police observation in Irkutsk, Siberia. He was implicated in the 1905 effort by revolutionaries to declare a separate democratic "Irkutsk Republic." However, whether Palchinsky was an active participant in the movement or just a sympathizer is not clear. At first the tsarist government saw him as a leader, and accused him of violating article 103 of the tsarist criminal code, which outlawed an "attempt to change the form of government in Russia." Evidently fearing that this charge could not be made to stick, the government revised its accusation to violation of article 126, which banned "participation in an organization known to have the goal of overthrowing the existing governmental order."[21] The difference seems to be that the authorities believed they could prove that Palchinsky associated with revolutionaries, but doubted that they could prove he actually participated in an attempt to overthrow the government. Palchinsky sometimes attended meetings organized by the anarchists or Socialist Revolutionaries, but there is no evidence in his archive that he was a formal member of any political party or revolutionary organization. He was not brought to trial in 1905, but simply sent into exile under the emergency powers granted to the police during the revolutionary turmoil.

Both before and after the 1905 Revolution Palchinsky rejected violence as a political means. As time went on, he became more and more interested in the Socialist Revolutionary Party, which from 1905 to the time of the Bolshevik Revolution in 1917 was the largest party in Russia. Among the Socialist Revolutionaries, he sympathized with the moderate wing of the party and was sharply critical of the radicals who promoted the assassination of tsarist officials to bring about a change in the regime. During the tenure of the Provisional Government in 1917 Palchinsky opposed those anarchists who joined with the Bolsheviks in trying to bring down the moderate socialist government.

While in exile under police supervision in Siberia in 1906 and 1907, Palchinsky continued to work as an engineer and became an expert consultant on mining operations. He was valued by mine owners for his ability to improve productivity and reconcile differences between management and labor. Despite his successes as an engineer, Palchinsky objected to the police controls, and in August 1907, he escaped Siberia and returned to Ukraine, where he roamed from city to city to elude the authorities. His radical friends from the days when he had studied labor conditions in the Don Basin helped him by providing shelter. In early 1908 he managed to slip across the border and began a new life in Western Europe that lasted for five years. Meanwhile, his wife moved back and forth between St. Petersburg and Irkutsk, all the while pleading unsuccessfully with the authorities to dismiss the criminal charges against her husband. In 1909 Nina Aleksandrovna and her mother, Mariia Aleksandrovna Bobrishcheva-Pushkina, moved to Western Europe to be with Peter.

In Germany, France, England, the Netherlands, and Italy, Palchinsky became a successful industrial consultant and, perhaps more important, developed an approach to technical problems that he followed for the rest of his working life. He insisted on viewing engineering plans within their political, social, and economic contexts. One of his biggest assignments was to world seaports such as Amsterdam, London, and Hamburg; he later wrote a four-volume study on the ports of Europe that was published in several languages.[22] Palchinsky's employers asked him to improve the productivity and efficiency of these seaports, and he advised them that one could not unload or load ships effectively unless the workers had the skills and commitment to do the job. Improving seaport operations was not simply a matter of providing cranes, rail spurs, deep sea channels, wharves, and warehouses; it was also a matter of workers' housing, schools, public transportation, medical care, recreational facilities, adequate pay, and social insurance. He viewed each seaport as a vast interlocking system of services that would permit workers to "achieve the maximum result with the least effort."[23] He was what the American historian of technology Thomas Hughes would call a believer in "technological systems."[24] To Palchinsky a seaport was similar to a large mine in that both required the movement of voluminous materials over considerable distances. The various parts of the operation must mesh as perfectly as possible, which meant that both the technology and the workforce must be in optimal condition.

Palchinsky seems to have adjusted well to life in Western Europe, where he learned several new languages. He maintained his contacts with Russia, however, and in article after article he advised the tsarist government that had forced him to flee how to improve its industry.[25] The obstacles to Russia's industrial advancement were not technological, he believed, but political, social, legal, and educational. He was convinced that the mineral wealth of Russia had fated it for industrial greatness, if only a government could be created that favored rather than feared the social effects of modernization. The legal system must be reformed to bring order to land titles, which, he wrote, were currently so disorganized that railroads and mines were almost impossible to build because no one knew who owned the land.[26]

Palchinsky was particularly critical of engineering education in tsarist Russia. He believed that the Russian engineering curriculum was too heavy in natural science, mathematics, and "descriptive technology," and almost totally ignored subjects such as economics and political economy.[27] Thus, he continued, graduates of Russian engineering schools think that every problem is a purely technical one, and they assume that any solution that incorporates the latest science is the best solution. No wonder Russian engineers are unequipped to deal with the competitive world—and that Russian technology cannot compete in the world market even though it is protected by high tariffs. He urged Russian engineers to stop approaching problems in what he called an "academic-dilettantish" way and instead to become hard-headed, realistic engineers who evaluate problems in all their aspects, particularly the economic ones.

In 1911 Palchinsky organized a Russian manufacturing and mining exhibit at an international trade fair in Turin, Italy, for which he received a special award from the Italian government.[28] Palchinsky was convinced that Russia could sell coal and ores on the world market if it would only take the necessary political and economic steps. His posture in the photograph taken at the Turin exhibit— arms and legs akimbo—conveys his pride and self-assurance.

Portrait of a Marriage

Peter Palchinsky's wife Nina was almost as active as her husband, matching his interest in industry and trade with her own in workers' education and the status of women. Before leaving Russia she taught in special schools for workers in St. Petersburg, Irkutsk, and the Don Basin, where she not only helped her pupils achieve literacy but taught them political doctrines of reform and change. In England, France, and Belgium she studied the women's struggle for voting rights and higher education; she wrote articles about these movements for a feminist journal in St. Petersburg entitled *The League of Women.*

Nina's family had historically favored reform or even revolution. Two of her ancestors, N. S. Bobrishchev-Pushkin and P. S. Bobrishchev-Pushkin, had been Decembrists, rebel officers who in 1825 unsuccessfully attempted to force the tsar to create a constitutional government. After their arrests the first was exiled to Siberia for life, the second for twelve years. Nina Aleksandrovna's father, Aleksandr Mikhailovich Bobrishchev-Pushkin, was a writer and a jurist who served as chairman of the St. Petersburg circuit court and as legal consultant to the tsarist government. He promoted reform of the legal system and emphasized freedom of conscience, especially on religious questions.[29] His poems, published after his death, revealed his deep disappointment in a legal career that had been regarded by many as successful.[30]

Not surprisingly in the context of late imperial Russia, Nina and her husband Peter were at first considerably more radical than their parents. They were critical not only of capitalist economics but also of bourgeois social relations. They believed that the typical marriage in capitalist society was stultifying and self-centered. A person should justify his or her life not by finding private happiness in a spouse, or in sex, or in children, or in comfortable living, but by doing good in the society at large.

Peter and Nina often had difficulties reconciling their radical social and political views with their private feelings, which were more conventional than either wanted to admit. Nina's desire not to have children was painful to Peter, despite the fact that both had agreed that children usually caused the family to turn inward from social concerns, a consequence they deplored. Peter claimed that he believed in equality in marriage, but his mode of life in the early years of their marriage usually cast Nina into the traditional position of maintaining a refuge for a husband constantly on the move. His years in administrative exile or prison consolidated Nina's role as the steadfast supporter—the person who brought presents and food to the prison or place of exile, and who beseeched the authorities for alleviation of his conditions or commutation of his punishment.

Peter and Nina believed that between husband and wife sexual relations were less important than political and social solidarity. Nina admitted on at least one occasion that she sometimes looked upon sex as a disruptive, even degrading, activity. In a February 1908 letter from St. Petersburg to Peter in Europe she wrote that she had been reading Tolstoy:[31]

I have just read again, thoughtfully, the "Kreutzer Sonata" . . . The way life is "arranged" would certainly be great if everything turned out as he says: marriage would be a spiritual arrangement, i.e., a spiritual closeness and friendship, a living comradeship; physical relations would be restrained in frequency, and only for reproduction. Such relations would not be elevated to a means of pleasure, but, on the contrary, they would be something to be ashamed of and would be looked upon as something that denigrates human beings. And of course they would be permitted only to husband and wife, i.e., only with one other person, and would not be permitted with others. How pure it would be to live in such a world!

Although the ideal of a Platonic, loving relationship between husband and wife and their abstinence from sex with others appealed to Nina on an abstract level, she recognized that she and Peter were far from achieving such a goal. Peter had admitted to her that in his travels he sometimes found sexual solace elsewhere. Without being so explicit, Nina in turn confessed to Peter that she was attracted to other men. She regretted her own and her husband's fallibility but, in the final analysis, maintained that it was not very important. Far more significant was a mutual commitment to the benefit of society as a whole, honesty with each other, and a "communion of souls."

In the years 1909 to 1913, when both Peter and Nina were in Western Europe, they often lived apart, each pursuing his or her own concerns. Nina and her mother lived usually in Turin or Genoa and visited frequently with the Kropotkins in Geneva, with whom conversation generally revolved around socialism and Russia. Like her husband, Nina traveled frequently, following her interests in education, art, and the women's movement, while Peter pursued mining, industry, and the management of maritime ports. Their occupations naturally took them along different travel routes, and they considered such divergence normal for two independent people. Yet they corresponded almost constantly, often in Italian, and they invented terms of endearment for each other in several languages.

Occasionally they would join up for grand adventures in Venice, Milan, Rome, or Paris. Then they would pose like teenagers in front of the standard tourist attractions, snapping pictures of each other and asking friends or passers-by to take photographs of them together, arm in arm.

Nina's belief that sex was too insignificant to cause marital difficulties received a test in December 1909, when Peter visited for three days with Vita and Alfred Shenk, who lived in Vienna with their two children, Iurii and Olga. Vita and Peter had had a romance many years earlier, before Peter married Nina. In contrast to Peter and Nina, Vita and Alfred were living a conventional life, doting on their children and enjoying the many bourgeois pleasures of early twentieth-century Vienna. Yet they seemed unhappy with each other. Vita, in particular, expressed dissatisfaction with the closed atmosphere of their home. In a letter written on the train after his departure from the Shenks, Peter described the family to Nina, and he also confessed that his old attraction to Vita had been instantly renewed. Indeed, he admitted that during his brief stay in Vienna he and Vita again became lovers.

At first Nina, true to her ideology, responded with affection and understanding:[32]

> My dear and good Petik, today I received your letter from Vienna, full of Vita. You are a darling that you speak so openly with your wife-friend, and it is very good that we can talk with each other, understanding everything simply and correctly . . . Far be it from me to be jealous of you, and never will I speak nor even think a word of reproach. Isn't it true that we are, first of all, human beings, and friends with each other, and, moreover, a man and a woman? I do not conceal from you that also my thoughts are often occupied with other people, but my liking for them does not in the slightest prevent me from loving my Petik more than any-

one else on earth or prevent me from feeling that I am his indivisible other half toward whom alone I experience passionate feelings . . . Why do the Shenks watch over their children so? Are they sick? What a purely bourgeois life they have! When I read your letter about them, I felt simply suffocated, just as if my head was in a bag. How far their life is from that path upon which we are traveling, and, especially the one on which you are now going, leading toward anarchism. Our life is that of the broad world, all of humanity, and theirs is a narrow family life with petty-bourgeois interests.

As the weeks passed and Nina continued to think about Peter and Vita, understanding and tolerance were gradually replaced with jealousy. The aspect of Peter's affair with Vita that particularly upset Nina was that it was a renewal of a bond that had existed before she and Peter were married and therefore seemed to be something more than a casual physical encounter. It threatened the "communion of souls" that was the basis of her relationship with Peter. An attraction between Peter and another woman that could survive ten years of separation was obviously a serious matter.

A little more than a month after she wrote the first letter, she wrote another, this one from Ghent:[33]

Your letter from the train about your stay in Vienna imposed a load on me that, to my regret, turned out to exceed my strengths as a *woman*. In *that* way I cannot share my Petik, especially his soul. I cannot accept this sharing without pain, but I will submit to it if absolutely necessary . . . In the very darkest moments it seemed to me that all your love for me was a mistake, that you really have loved only one person, that you have always loved one and the same woman, and therefore in three days that woman could again possess your soul, that I have completely retreated into the background.

Nina's initial tolerance followed by an expression of overt jealousy may well have been the wisest way for her to save their marriage. Peter resumed his loving correspondence with Nina, and mention of the episode disappeared from their letters. But they continued to maintain that they were free of bourgeois attitudes toward romance and sex. They remained friends of Vita Shenk and her husband Alfred for many years. The marriage of Nina and Peter continued, at least to all appearances, to be an affectionate comradeship cemented by a common interest in social and political issues.

Return to Russia

In 1913, when his eight-year Siberian exile would have ended had he remained in Russia, and on the tercentenary of the Romanov dynasty, Palchinsky received a pardon from the tsarist government and he and Nina returned to their native land. There he established in 1916 an institute devoted to the "study of the rational use of the natural resources" of Russia. Known as the Institute of the Surface and Depths of the Earth, it took as a slogan a thousand-year old phrase from the primary chronicle of ancient Kiev: "Our land is great and rich, but there is no order in it." Palchinsky announced that his goal was to achieve order not by inviting in foreigners, as the ancient Kievans had done with the Vikings, but by applying the methods of modern engineering to problems of economic development. The institute began publishing a journal, *The Surface and Depths of the Earth*, which printed articles on mining and industry.

Like many people, as Palchinsky grew older and was recognized for his achievements he became more conservative politically, economically, and personally. He served on the board of a mining company and established close connections with the business community.[34] During World War I he was an advisor to defense industries and served as deputy chairman of the government's War Industry Committee.[35] In the latter position he began to see that centralized planning of industry, at least during wartime, had definite virtues. No longer did Kropotkin's emphasis on decentralization of economic activities seem as attractive to him as it once had, although he continued to admire Kropotkin, especially for his views on how technology could be made to serve the public interest. He considered himself a democratic socialist, and he favored the overthrow of the tsarist government for which he worked.

Palchinsky was a strong supporter of the Provisional Government that was established in Russia in February 1917, after the downfall of the monarchy. He saw it as providing the best possible opportunity for the emergence of democratic government in Russia. While probably not a formal member, he associated himself with the right wing of the Socialist Revolutionary Party, and supported the war effort against Germany. He held several positions in the Provisional Government, including deputy minister of trade and industry and, briefly, assistant to the governor-general of Petrograd. In his work in the Ministry of Trade and Industry he offended many members of the political left by supporting price and wage controls in industry, which he defended as emergency measures required by the military situation.[36]

According to Bolshevik mythology, the takeover on the night of October 25, 1917, of the Winter Palace, where top officials of the Provisional Government had sought refuge, was a heroic military event. In actual fact, the Winter Palace fell by infiltration rather than by frontal assault, and only a few people were killed.[37] The day before the palace was taken, the ministers met and appointed Nikolai Kishkin, the minister of welfare, governor-general of St. Petersburg; Kishkin in turn named two engineers, Peter Palchinsky and Peter Rutenberg, as his assistants. These three men were supposed to organize the defense of the Winter Palace.[38]

Palchinsky later wrote that he was horrified by the lack of discipline and resolve of the members of the Provisional Government, and soon recognized the situation as hopeless.[39] Nonetheless, he organized a "defense" as best he could, and even arrested several Bolsheviks whom he found wandering around the palace corridors, thereby earning opprobrium in later Party accounts of the Bolshevik revolution. Palchinsky believed fervently that the Provisional Government was a legitimate one that should yield power only to a democratically elected alternative, which he saw in the coming Constituent Assembly. In the end, however, as the Bolshevik soldiers entered the palace's Malachite Chamber, where the ministers were seated around the table awaiting their captors, Palchinsky ordered the few remaining loyal guards not to shoot.[40] The ministers and other officials of the Provisional Government, including Palchinsky, were taken prisoner.

From Political Prisoner to Soviet Consultant

In the first hours after their seizure by the Red Guards in the middle of the night the lives of Palchinsky and the other officers of the Provisional Government were in danger. One of the Bolshevik supporters who rushed into the Malachite Chamber, enraged on seeing that the head of the government, Aleksandr Kerensky, had escaped, shouted, "Bayonet all the sons of bitches on the spot!"[1] But the Bolshevik leader of the armed detachment, Vladimir Antonov-Ovseenko, quieted the group by affirming, "The members of the Provisional Government are under arrest. They will be confined to the Peter-Paul Fortress. I will not allow any violence against them."[2] He ordered that the prisoners be escorted out of the Winter Palace and across the bridge over the Neva River to the prison fortress a short distance away.

As soon as the group emerged onto the dark street it was surrounded by a fist-shaking mob of Bolshevik adherents who demanded that the government officials be beheaded and thrown into the river. At that moment gunfire broke out from several different directions. The Bolshevik soldiers and sailors occupying the Peter-Paul Fortress, seeing a crowd headed toward them, thought they were under attack and responded with volleys of machine-gun fire. In the general panic that erupted both the government ministers and the members of the crowd who threatened them ran for cover. Somehow the sailors and Red Guards who had been ordered to escort the prisoners to the fortress managed to collect them and hurriedly pushed them across the bridge to the safety of the fortress.[3]

Inside, Palchinsky and the other officials were counted, their names were taken, and they were led to cells in the Trubetskoi Bastion, an ancient section of the fortress that had held generations of Russian dissidents (including, many years earlier, Palchinsky's mentor Peter Kropotkin). There Palchinsky joined a mixed lot of other prisoners, some of whom had been locked up before the overthrow of the monarchy months earlier. For the next four months Palchinsky lived with these men, whose political loyalties were heterogeneously linked to socialism, monarchy, and liberal democracy. Among the best known of the prisoners were Mikhail Tereshchenko, a Ukrainian businessman and former minister of finance and foreign affairs in the Provisional Government; Vladimir Purishkevich, a leader of the reactionary faction in the pre-Revolutionary legislature, the Duma; Pitirim Sorokin, an eminent sociologist who later became a professor at Harvard University; Vladimir Sukhomlinov, minister of war in the tsar's government from 1909 to 1916; and Fedor Kokoshkin and Andrei Shingarev, leading members of the Constitutional Democratic Party.

The prisoners were confined in small, cold, and dirty cells, each with one heavily barred window.[4] The daily regime was strict, but not inhumane. Up at 7:00 a.m., the prisoners received hot water, a little sugar, and a quarter pound of bread. At noon they were given hot water, some cabbage, and a bit of meat. At 4:00 in the afternoon, they received afternoon tea. At 7:00 p.m. they received more hot water in place of dinner. From 8 until 10 in the evening the prisoners were permitted to mingle and exchange rumors. Their cells were wired for electricity, but current flowed only about an hour a day. Mail was permitted daily and visitors weekly. Nina Aleksandrovna visited Peter Palchinsky regularly, and took away the articles that he continued to write. One of them, on the restoration of economic life in Russia, was actually published, although someone deleted Palchinsky's reference to the "view from his cell window."[5]

Palchinsky and his fellow prisoners were permitted to attend church in the fortress cathedral, which was known throughout Russia as the burial place of the tsars. Some of the prisoners were not religious, but the trip to the cathedral, where they could stand among the tombs of the most famous figures in Russian history, was an event they all eagerly anticipated. Several of them knew Russian history well, and they entertained their prison mates with details from the careers, often gory and venal, of the past rulers of their land.

At first Palchinsky's mood in prison was quite positive. He had seen the inside of jail cells before, as had a number of his comrades, and he was certain they would survive. He was elected *starosta*, or leader, of a group of prisoners who lived in close proximity in the fortress cells (Kishkin, Tereshchenko, Rutenberg, Avksent'ev, Sorokin, and Shmelev). On October 28, 1917, he wrote Nina Aleksandrovna: "Be calm and do not worry. As you know, the worse things get the calmer and better I feel . . . Please do not make any special efforts on my behalf."[6] Two days later he observed, "After the dog's life of the last 8 months I am resting and am even pleased that no one is nearby." He included in his letter a drawing of his cell, no. 43 in the Alekseevskii *ravelin*, a detached fortification with two embankments jutting out from the bastion. Ten days later he wrote his wife that the jailers were permitting them to read some newspapers, especially the Bolshevik party paper *Pravda*. Knowing that his letters were read by censors, he nonetheless described *Pravda* as "full of lies," and joked about the "dirty hands" through which his letters passed before reaching his wife.[7]

The situation of the prisoners soon began to deteriorate. Commandant Pavlov, the head of the prison, warned his charges that militant soldiers and sailors were threatening to break into the fortress and murder the "leaders of the old regime" allegedly ensconced there. Indeed, two of Palchinsky's prison mates, Kokoshkin and Shingarev, were murdered in this way after they had been transferred to the prison hospital for treatment of tuberculosis, where they were more accessible to the militant revolutionaries. After news of this grisly event, the morale of the prisoners plummeted.

Release from Prison

By early 1918 the Bolshevik government was taking an increasingly tolerant attitude toward "bourgeois specialists," whose help was needed in the economy and in the Civil War. Palchinsky was released from Peter-Paul Fortress on March 7, 1918. He had not been out of prison more than three months when he was arrested again, on June 25, 1918, without any specific charges being made against him. This time he was held for almost nine months. While Palchinsky was in prison a group of Left Socialist Revolutionaries, members of a terrorist party with which Palchinsky strongly disagreed, made an attempt on Lenin's life; in retaliation, the Soviet government announced that 122 prominent prisoners, including Palchinsky, would be shot if any Soviet officials were killed.

Palchinsky once again escaped death, however, bouncing back like the Russian toy his half-sister Julia had described nine years earlier. A Swiss Social Democrat named Karl Moor, who knew Palchinsky's technical work in Europe, wrote to Lenin urging that he be released. In turn, Lenin wrote to Grigorii Zinoviev, the head of the Petrograd section of the secret police:[8]

> Comrade Karl Moor, a Swiss, has sent me a long letter asking for Palchinsky to be set free on the grounds that he is a prominent technician and organizer, author of many books, etc. I have heard and read about Palchinsky as having been a speculator, etc., during Kerensky's time.
>
> But I do not know whether there is now any evidence

against Palchinsky? Of what kind? Is it serious? Why has the amnesty law not been applied to him?

If he is a scientist, a writer, could he not—if there are serious charges against him—be given special treatment (for example, house arrest, a laboratory, etc.).

There were no serious charges against Palchinsky, and his talents fitted him well with the new Soviet policy of employing specialists from the old regime in industrial tasks. Palchinsky was again released from prison on March 17, 1919. For the next eight or nine months he feared being rearrested, and stayed with friends (I. M. Gubkin and L. T. Rabinovich) in Moscow rather than in Petrograd, where his wife continued to reside.

At first Palchinsky, like the great majority of technical specialists in Russia immediately after the Revolution, had little sympathy with the Bolsheviks, who were, in his mind, usurpers of power. Gradually, however, he and many of his associates found certain aspects of the new Soviet economic and political system beckoning. The Bolsheviks were committed to creating a planned economy, to industrialization, and to science and technology. They seemed eager to benefit from the services of engineers and scientists. Long before the Revolution Palchinsky had been a socialist, and during World War I he had become accustomed, as an administrator of war industries, to the idea of a command economy. Perhaps he could work with the new rulers of Russia after all.

He volunteered to help the new planning agencies that proliferated immediately after the Bolshevik victory, and soon was very busy. Within a few weeks of leaving prison in mid-March 1919, he was consulting for a variety of Soviet offices. It is ironic that he began his work for the Bolshevik government at the same time that he was dodging the police by staying away from his home in Petrograd. Nina Aleksandrovna wrote him frequently from Petrograd and advised him to keep a low profile, a suggestion that he ignored; he maintained constant contact with Moscow economic and military authorities.

On April 23, 1919, Nina Aleksandrovna wrote that the Cheka (secret police) had recently arrested one of their acquaintances and urged him "not even to think about returning to Petrograd."[9] She later wrote to report a rumor that she had heard. According to several of her informants, Lenin in a meeting of the Central Committee of the Party had recommended that Palchinsky be named commissar of trade and industry, but had backed down in the face of objections; one protester asked ironically, "Why not Miliukov?"—leader of the liberals, foreign minister in the Provisional Government, and famous opponent of the Bolsheviks.[10] While it is a delightful thought that at a time when Palchinsky was eluding the police the leader of the Bolsheviks was considering him for a high position in the Soviet government, the story is probably not true. Lenin valued Palchinsky's technical abilities, but thought that he could not be trusted politically and therefore probably would not have recommended him for such a responsible position.[11] Such contradictory rumors about Palchinsky illustrate the confusion of the times.

On June 18, 1919, Nina Aleksandrovna reported a more ominous story to Peter. The police had come by their apartment building in Petrograd and asked, "Does Palchinsky live here?" A neighbor had replied, "He hasn't lived here for a long time." Nina Aleksandrovna added in her letter to Peter,[12]

> Just who these police were—the local ones or the secret police—is unknown. But the damned fellows have not forgotten about you.
>
> It is clear that you can't live in Peter[sburg] for a long time yet. The fact that you are working in Moscow in so many different places is, in my opinion, both good and bad. Best of all would be if your name was not mentioned anywhere or

to anybody. But at the same time your good work for them means they can't reproach you for sabotage.

A little later in the same letter, Nina observed that she was reading Romain Rolland's *Jean Christophe* and was struck by how similar the hero of that novel was to Peter: "Here are some phrases that describe both him and you: 'He had a stubborn desire for life, for action . . . It would be better to live a full life and burn up quickly than to save oneself.' And again, 'He was one of those people who wanted to take action up to the moment when there was nothing left to do, up to the last possibility.' I see you in just this way."

Consultant to the Soviet Government

Palchinsky was, indeed, taking action up to the last possibility, striving to unite his ideas about industrial planning with Bolshevik aspirations. He was particularly excited by the plan to electrify all of Russia in a few years, and accepted the idea, quite popular at the time, that a centrally planned socialist economy would be able to electrify much more rapidly than a capitalist one.[13] He became a professor at the Mining Institute and worked as a consultant on a host of projects, including the building of the giant dam on the Dnieper River, the drafting of maps of population density and mineral deposits, the building of railroads and mines, and the construction of sea and river ports.[14] Because of his abilities he was sought out by many government planning agencies. He quickly became one of the best-known engineers in Soviet Russia, serving as chairman of the Russian Technical Society and a member of the governing presidium of the All-Russian Association of Engineers. Throughout this period he was amazingly active, writing dozens of articles and reports for government committees.

In 1922 he renewed his youthful anarchist connections by attempting to honor publicly the memory of Kropotkin, after which he was once more jailed, this time for two months.[15] Again he was rescued by a high official, this time by Gleb Krzhizhanovskii, the chairman of the State Planning Commission (Gosplan), who was eagerly awaiting a report he had commissioned from Palchinsky on the metals industry. Palchinsky had continued to work on his consulting assignment while in prison. On January 16, 1922, two days before the report was due, Krzhizhanovskii wrote to the Moscow police authorities:[16] "Considering that the permanent consultant of the State Planning Commission, Engineer P. A. Palchinsky, is to deliver a report in the Southern Bureau on Jan. 18 of this year at 3 p.m. on the question of restoring Southern Metallurgy, which is of particular significance at the present moment, the Presidium of the State Planning Commission requests the Revtribunal [Revolutionary Tribunal] to release Comrade Palchinsky by the above-named hour so that he may carry out the assignment given him."

For the third time Palchinsky was released from Soviet prison, and went almost immediately to his industry consulting assignment. He quickly threw himself into his work, and was soon again one of the most sought-after engineers in Russia. He traveled constantly, wrote many reports for government commissions, and everywhere championed the causes of mining and industry.

In the early 1920s one of Palchinsky's friends was Maurice Laserson, a French financial expert who was helping organize the State Bank. Laserson reported that at a meeting in May 1923 he asked Palchinsky if he were not afraid that he would be arrested again. Palchinsky replied, "I remain here because I am eager to work here. This is my place. I do not believe that I have anything to fear after all that I have already gone through. I am no longer fighting them, why should they do away with me? And if my hour comes, well you know the Russian proverb: 'You can't have two deaths, but then you can't avoid one.'" According to Laserson, Palchinsky spoke "of the necessity and the duty incumbent upon every intellectual Russian—now that the Soviet Regime had apparently turned from destruction to reconstruction—to serve his country, however much he might hate or despise this regime of violence."[17]

In his effort to promote mining in the Soviet Union, Palchinsky even wrote poems that he read at festive meetings of professional engineers. Without the rhyme, which is lost in translation, I will cite part of one poem, presented at a banquet for engineers of the Mining Institute in 1925:[18]

> We believe in the power of science,
> Walking vigorously through life,
> And we hope that our descendants
> Will remember us with kindness,
> Those who put their labor in mining.
> What was burning inside us for a long time,
> What we suffered from being oppressed,
> What had forced us to strain our strength,
> Let this all in this speech be expressed,
> And approved by the ringing of glasses.
> Let the business of mining be growing,
> Let the Institute blossom and brighten,
> Let our merger and labor be stronger,
> Labor intelligent, powerful, skillful.
> I drink to the prosperity of mining!

Conflicts with the Communist Party

Palchinsky was eager to work with the Soviet authorities and the Communist Party in planning industry and increasing the strength of Russia, but he stoutly resisted the takeover by the Party of any organization of which he was a member. He made a sharp distinction between the interests of Russia as a country, which he strongly supported, and the interests of the Communist Party, which he opposed. When, in December 1924, the All-Russian Association of

Engineers, of which he was an officer, was forced to accept the Party's control, including a new slate of pliant leaders, he resigned from the association.[19] When later asked if he would rejoin, he replied, "To my regret, there can be no talk of my return . . . so long as the association does not feel that it is again a free engineering organization and does not divest itself of those leaders who have been thrust upon it and who have deprived it of its character."[20]

Convinced that Party members were intent on taking over any organization they were allowed to join, Palchinsky worked to keep them out of his Institute of the Surface and Depths of the Earth. When the prominent geologist and petroleum engineer I. M. Gubkin, who was a member of the institute, indicated in 1921 that he might join the Party, Palchinsky wrote him that such a step "would be of interest to the Institute."[21] Gubkin, an old friend of Palchinsky, stiffly replied that if he decided to enter the Party he would do it openly and take full responsibility for it. At the same time, Gubkin resigned from the institute, achieving Palchinsky's goal of keeping it free from Party influence. Gubkin shortly afterward did, indeed, join the Communist Party and was active in creating a system of Party control in several other organizations, including the Academy of Sciences, to which he was one of the first Communists elected to membership.[22]

Palchinsky's outspoken ways often got him in trouble. In the early twenties he was appointed a permanent member of Gosplan and regularly attended its meetings. However, in February 1924, the chairman of Gosplan, G. M. Krzhizhanovskii, became unhappy with Palchinsky's frequent criticism of Party policy. Hearing of Krzhizhanovskii's irritation, Palchinsky wrote him a letter resigning his membership in Gosplan.[23] Krzhizhanovskii was no doubt relieved to have this gadfly outside his formal organization, but at the same time recognized his talents by urging him to continue working with Gosplan on a consulting basis, which Palchinsky did.

In 1926 Palchinsky made a 12,000-kilometer trip through

Soviet Central Asia, traveling for sixty-five days by train, steamboat, horseback, and foot. His assignment from the Soviet government was to evaluate the potential of the oil and gas industry. He was as willing to criticize Soviet industrial practices as he had been to criticize tsarist ones. In particular, he castigated what he called the "gusher psychology" of the administrators of the oil industry, who wanted to drill wells that would produce spectacularly and impress the top officials in Moscow, meanwhile ignoring large amounts of coal and gas that were often more economical sources of energy.[24] Palchinsky remained skeptical of commands from the center that ignored local conditions, just as he had been before the Revolution, when he criticized the tsarist government for importing from Western Europe stone for foundations and embankments that could have been obtained near at hand.[25] And he continued to defend workers against managers. He wrote in 1927 that the Soviet oil industry has "too many administrative rules and too few safety rules."[26] He believed that the managers of oil refineries talked too much about preventing theft and hooliganism and too little about protecting the workers from fires and explosions. Theft and hooliganism, he said, were violations that should be covered by criminal laws. The special obligation of plant administrators and construction supervisors was to protect the lives of the workers.[27]

Palchinsky was an independent and even stubborn man who refused to give an evaluation of a problem before he had patiently collected all the relevant data. In 1928, after the start of the First Five-Year Plan, with its emphasis on accelerated productivity, the Supreme Economic Council (VSNKh) asked Palchinsky to make recommendations for the location of new coal mines in the Cheliabinsk region.[28] Palchinsky went about the task with his usual thoroughness, requesting voluminous data on coal deposits, transportation systems, and population density. In April he was asked to hurry up his work to meet the strict timetable laid down by higher authorities. Palchinsky coolly replied that all the data that he had requested were not yet in, and that some of the data were essential for the evaluation, and therefore he could not hurry the completion of the project.[29] His only concession was to postpone his request for his own pay.

Independent Planner of Soviet Industry

In the 1920s Palchinsky developed his own program for industrializing the Soviet Union. He expanded the Institute of the Surface and Depths of the Earth that he had established in 1916, promoted its journal, gathered a group of similarly inclined engineers around him, and wrote a steady stream of editorials and articles calling for the planned advancement of Soviet industry, which would rely heavily on the unparalleled mineral riches in the lands of the Soviet Union.[30] In February 1922 he organized in Moscow the Miners' Club, a voluntary organization with the goal of independently evaluating mining projects and issuing periodic reports. Keeping up with the foreign literature in his field, he read and reviewed works in English, French, German, and Italian.[31] Palchinsky, however, did not merely copy the ideas of foreign writers, but instead developed conceptions particularly applicable to Russia. He emphasized that the Revolution had overcome many of the obstacles to industrialization that engineers had earlier encountered. He believed that the new Soviet regime presented possibilities for the planning of industry about which engineers of the tsarist period could not have dreamed; indeed, he thought that Soviet engineers, freed from capitalist employers, could have a greater influence on their nation than engineers anywhere else. He hoped that Soviet engineers might come to occupy the roles that financiers and entrepreneurs had under capitalism.[32]

Although Palchinsky praised the idea of central planning, he thought that the central plan should be very general, allowing many local variations. It should allow room for individual initiative. Conditions on the spot, such as availability and costs of coal, water transport, educated workers, and construction materials, would lead to different solutions to problems that at first glance appeared similar.[33] Whether to use wood or coal as fuel for steam engines on Soviet railroads should not be dictated from Moscow, he wrote; instead, fuel should be obtained locally according to price.

Soviet economic planners of the late twenties were divided between supporters of "functional" planning and supporters of "regional" planning. Functional planners believed that one should plan from the center, Moscow, and think in terms of entire branches of the economy, such as the steel industry, without much regard for local differences. Regional planners believed that planning should grow "from the ground up," and should be based on careful studies of the characteristics of local areas: mineral deposits, population centers, transportation networks, cultural and recreational facilities, and so forth. Palchinsky favored a combination of the two, but in his reports his greatest emphasis was on regional planning. His attention to local conditions was an effort to counteract the tendency of planners in Moscow, especially as the Communist leaders concentrated more and more power in the capital city, to ignore them. Palchinsky believed that one should go out into the provinces, study the existing situation—not only the natural resources but the human ones as well—and then plan industry in a way that allowed for many variations based on the characteristics of specific localities. He was impatient with Gosplan officials who accused him of "parallelism" when he called for both regional and functional studies. It was the job of Gosplan, he insisted, to combine these approaches.

Part of Palchinsky's emphasis on local differences came from his education in mining engineering, a field that is very site-specific, but he also believed that blind, centralized planning was inefficient and unjust. It not only directed the development of industry in a less productive way, but also overrode the local populations and ignored their specific situations and needs.

Palchinsky, writing in 1922, criticized sharply the Bolshevik penchant for gigantic enterprises—the belief of Party leaders that the best facilities will always be the largest ones. He lamented the ideology that considered small industry, workshops, handicraft workers, and artisans to be relics of the past. He asked rhetorically, "Is it possible to build locomotives, oceangoing ships, bridges, and gigantic hydraulic presses in small workshops or handicraft centers? Of course not. But do we need gigantic factories to have good buttons, good socks, office materials, tableware, clothing, etc.? Of course not."[34] He made a plea for a symbiotic mixture of all kinds of industries and workshops. The Soviet Union, he warned, must have a goal beyond the construction of heavy industry. It also must aspire to a society where all human needs are fulfilled, a goal that cannot be achieved without heterogeneity in scale, style, and organization.

While in exile in Western Europe before the Revolution, Palchinsky had been impressed that even in heavy industry small-scale operations were sometimes the most efficient. He wrote in 1911, for example, that in the British coal industry the largest contribution was made not by the combined efforts of the largest mines (above 1000 workers) but by the combined efforts of middle-sized mines (100 to 1000 workers). The latter produced 70 percent of Britain's coal, while the former produced 28 percent.[35]

As an advisor to Soviet industrial planners after the Revolution, Palchinsky continued this line of reasoning. He noted that middle-sized and small enterprises often have advantages over large ones.[36] Replacing machinery is often easier in small facilities. Supervision is usually simpler and more intimate. Workers at small and middle-sized plants are usually more successful in grasping the final goals of the enterprise. Small plants, he concluded, have the psychological advantage that the entire staff usually feels organically interconnected.

Some plants will of necessity be large. Since Russia did not have the capital to initiate these facilities, Palchinsky recognized that the money would have to come from abroad. In 1922 he proposed that the All-Russian Association of Engineers act as a nonprofit advisor to a shareholding society that would bring foreign investments to Russia.[37] Later he dropped the idea and warned that foreign concessions should be "tightly controlled."[38] As a socialist he worried about foreign capital's becoming too important in the Soviet economy. Palchinsky's wary call for foreign investment fitted well with the New Economic Policy of the twenties, but it would be cause for criticism when the foreign concessions were cancelled at the end of the decade.

Humanitarian Engineering

The single most important factor in engineering decisions, Palchinsky maintained, was human beings.[39] Successful industrialization and high productivity were not possible, he repeatedly emphasized, without highly trained workers and adequate provision for their social and economic needs. An investment in education promoted industrialization more than an equivalent investment in technical equipment, since an uneducated or unhappy worker would soon make the equipment useless.[40] Simply adding new equipment without attention to the morale and ability of the workers would cause great waste. Human beings must be considered not hired hands but creative individuals with cultural and spiritual needs. Concern for satisfying the needs of the workers is not just an ethical principle but a requirement for efficient production.

Palchinsky observed in 1926 that although industry in Russia had recovered from the damage suffered in war and revolution, some enterprises were far more efficient than others.[41] These differences were not rooted in the availability of equipment and technology. Some of the most productive industries were less well equipped than the less productive ones.[42] The difference came down to the workers—how well educated were they, how well treated were they, how interested were they in the work.[43] Concern for human beings, Palchinsky advised Russian managers, "will bring more fruits than any-thing else."[44] The rebuilding and expansion of Russian industry must be based on "inner renewal," not achieved through the imposition of a forced pace from above or the importation of foreign tech-nology.

Palchinsky believed that socialist Russia had the opportunity to develop a far more humane industry than anywhere else. While Palchinsky admired American workers, for example, he thought in-dustrial managers in the United States were too narrowly interested in profits and that American society in general was too self-centered. In place of the Monroe Doctrine that "America is for Americans," Palchinsky proposed a new principle:"The world is for its human inhabitants."[45]

Paradoxically, Palchinsky and some of his non-Marxist col-leagues at the Institute of the Surface and Depths of the Earth were more critical of American methods of management than the Bolshe-vik leadership itself was. The Marxists who took over Russia in 1917 were eager to utilize the latest methods of industrial management, and they turned to the doctrines of the Americans Frederick Winslow Taylor and Henry Ford for instruction. Taylor in the first decades of the century had introduced a system of time-and-motion studies that revolutionized machine-shop practice. He also rationalized the use of machinery and tools by workers and the layout of the workplace. Henry Ford introduced the principle of production flow and the assembly line to the manufacture of automobiles. He put his system into practice at Highland Park, Michigan, in 1913, just four years before the Bolsheviks came to power. The application of Taylor's and Ford's methods in American factories improved productivity but also increased the time pressure under which employees labored.

"Fordizm" and "Taylorizm" became standard terms in the lexi-con of the Soviet industrializers.[46] Lenin himself sanctioned the new industrial management methods when he asserted in 1918, "We must introduce the Taylor system and the scientific American system of increasing the productivity of labor throughout all of Russia."[47] Aleksei Gastev, director of the Institute of Labor in the 1920s, declaimed to the Soviet workers, "Let us take the storm of the Revolution in Soviet Russia, unite it to the pulse of American life, and do our work like a chronometer."[48]

Palchinsky and several of his fellow engineers were all in favor of the gains in efficiency that assembly lines and stopwatches might bring, but they worried about the mind-numbing effects that the imposition of Taylor's and Ford's methods might have on the workers. Would the worker on an assembly line be regarded merely as a pair of hands, or a cog in the machinery, rather than an individual?[49] A socialist approach to rationalization (rather than a capitalist one) should, they thought, not merely emphasize time-and-motion studies, looking for the most efficient way to put a nut on a bolt, but should also improve workers' education and well-being. Workers who personally benefitted from improvements would be eager to find the most efficient way to assemble a product, and would work willingly with the management to this end.[50]

The more Palchinsky examined Taylorism the more critical of it he became, and he proposed "humanitarian engineering" as a substitute.[51] The essence of humanitarian engineering was upgrading the worker's knowledge to such a level that the primitive methods of Taylorism, designed for inexperienced workers, would not be necessary. Knowledgeable workers, Palchinsky believed, would be masters of their work, not slaves of it.

To achieve the goal of well-educated workers, Palchinsky proposed an expanded system of workers' schools, financed by the government but supervised by engineering societies, such as the Russian Technical Society. The Soviet government did not see any reason for the creation of special schools outside the normal educational system and was suspicious of the engineering societies, many of whose members, like Palchinsky, were not affiliated with the Communist Party. Undeterred, Palchinsky fought hard and unsuccessfully for his educational ideas, even writing in 1925 an appeal to the Party leader Leon Trotsky. (Palchinsky was unaware that Trotsky's political influence was already beginning to wane.)[52]

Ambitions for the Future

Palchinsky was a sophisticated man who approached technical problems in a remarkably broad fashion. He brought to these issues both the concerns for efficiency of an engineer and the concerns for social justice of the pre-Revolutionary political party, the Socialist Revolutionaries, with which he had sympathized. He even worried about the damage to the Russian forests that the excessive use of wood for construction and fuel might bring; he thought that a socialist government should be better able than a capitalist one to find a solution to such environmental problems.[53] Yet Palchinsky could also be hardheaded, willing to defend managers against what he saw as excessive environmental regulation. For example, he sharply opposed a draft project aimed at stopping wastage and environmental damage around mines and oil wells on the grounds that it was too restrictive and would make the job of supervisors and managers too difficult.[54]

Palchinsky promoted a very ambitious role for engineers. He wanted engineers to apply a new form of social analysis to problems of industrialization, and he believed that in order for this to happen the engineer's place in society must change. Earlier, the engineer had been assigned a passive role by society: higher authorities asked him to find solutions to technical problems. Now, Palchinsky maintained, the engineer must emerge as an active economic and industrial planner, suggesting where economic development should occur and what form it should take.[55] For example, an engineer asked to develop plans for a large hydroelectric dam on a major river should ask whether a dam was the optimal way to obtain electricity. What were the tradeoffs among the various methods of generating electricity? If coal were locally available, perhaps a thermoelectric plant would be a wiser choice. Answering the question depended on analyzing local factors and evaluating all the effects—economic, social, and environmental—of each option.

Palchinsky's vision of the new Soviet engineer was based on a justifiably strong conviction that a broad approach to engineering would result in more efficient industrial enterprises and more satisfied workers. The new model engineer also appealed to Palchinsky's professional pride. Edwin Layton observed that engineers in the United States in the same period displayed an "obsessive concern for social status."[56] Palchinsky and his colleagues were eager to promote the engineer to a new prominence in society, and they believed that the Soviet state, with its emphasis on centrally planned industrialization, provided unusual opportunities for this promotion.

For all his sophistication about engineering, however, Palchinsky badly misunderstood the political course of the Soviet Union. His ambitions for engineers could be realized only in a society that granted the various professions a high degree of autonomy and whose government was willing to listen to advice from outside official circles. As he was to discover, Stalin had a very different vision of society and of industrialization.

Stalinism Confronts Palchinsky

A historian with the advantage of hindsight could predict a collision between Palchinsky's program of industrialization and the one being adopted by the Communist Party. The tensions between Palchinsky's vision and that of the Party became extreme once Stalin gained absolute control at the end of the twenties. The primary issue was political authority. The Communists had never allowed professional groups to have the kind of autonomy or to express the broad concerns that Palchinsky wanted engineers to do.

Once Stalin had established control over the Soviet political and economic system, Palchinsky's proposals ran into even more problems. Palchinsky liked to say that a good engineer could not perform miracles, only the maximum within what was possible.[57] Stalin, by contrast, promoted an ideological campaign for economic advancement that set wildly unrealistic goals which required super-human effort. He insisted on the construction of gigantic hydro-electric power stations, which he found impressive in scale and revolutionary symbolism, regardless of the local conditions that Palchinsky found so important. Stalin demanded that industrial estab-lishments be of great size, preferably the largest in the world—an industrial policy that Western observers later characterized as "gi-gantomania." Palchinsky maintained that size was not, in itself, a virtue.[58] Stalin was quite willing to force poorly educated peasants from the countryside to perform tasks in new industries for which they were not qualified. The results were high accident rates and shoddy production, graphically described in memoirs of the period.[59] The relocated workers lacked adequate housing, especially for the winters. Their high death rate from exposure and disease was for Stalin an acceptable cost, but for Palchinsky it was a sign of irratio-nality, inefficiency, and injustice.

While Palchinsky called for moderation, saying "We are not magicians, we cannot do everything," Stalin maintained "There are no fortresses that Bolsheviks can not storm."[60] And while Palchinsky held that the human factor was of utmost importance in industrial-ization, Stalin stressed that "technology decides everything."[61] More than a little irony exists in the professional engineer's call for atten-tion to human needs over technology while the Party leader empha-sized technology above all else.

The most important source of conflict between the two was Stalin's mistrust of specialists educated before the Revolution. Stalin had served on a commission investigating the strikes of the university faculties and engineers immediately after the Revolution, and he considered the technical intelligentsia potential saboteurs.[62] In Sta-lin's eyes Palchinsky not only disagreed about how the Soviet Union should industrialize but also harbored dangerous ambitions. Palchinsky called for engineers to take an active role in politics; Stalin's opinion on this subject was revealed in an interview by H. G. Wells in 1934: "The engineer, the organizer of production, does not work as he would like to, but as he is ordered . . . It must not be thought that the technical intelligentsia can play an independent role."[63]

A sense of just how proud Palchinsky was of the engineering profession, and what a high role he assigned to science and technology, can be gained from a draft letter in the archive dated 5 December 1926. This letter was apparently intended for Aleksei Ivanovich Rykov, at that time the prime minister of the Soviet Union. In the letter Palchinsky asserts that science and technology are more important factors in shaping society than communism itself. This century, he writes, is not one of international communism, but of international technology. We need to recognize not a Komintern, but a "Tekhintern." His friends persuaded him, for good reason, not to mail the letter.[64]

Although Palchinsky was the engineer who most consistently advocated an alternative vision of industrialization and the role of the engineering profession, other engineers pursued similar concerns.[65] One center of technocratic thought in Russia in the twenties was the journal *Engineers' Herald* (*Vestnik inzhenerov*), edited by I. A. Kalinnikov, who had held many responsible positions in engineering education, including the presidency of the famous Moscow Higher Technical School.[66] Kalinnikov would later be accused, along with Palchinsky, of being one of the leaders of the Industrial Party. In 1927 Kalinnikov helped organize a discussion group, the Circle on General Questions of Technology, which announced its intention to develop "a whole new worldview, fully adapted to contemporary technical culture." One of the spokesmen of the Circle, the engineer P. K. Engelmeier, called for engineers to unite "not only along trade union lines, but on the basis of ideology."[67] Engelmeier's failure to mention Marxism in connection with this new ideology brought immediate criticism from Communist Party ideologues.

Another source of the Soviet technocratic movement was the technical advisors to the central Soviet economic planning apparatus. Under the Supreme Council of the National Economy (VSNKh) was a Scientific-Technical Administration responsible for developing policies guiding industrial research and development.[68] Palchinsky was one of its members. Most of its staff would later be brought to trial. These engineers called for the application of scientific methods not only to Soviet economic development, but also to such fields as industrial psychology and management. According to one of their documents, "the future belongs to managing-engineers and engineering-managers."[69] This was a phrase that Party critics later used against the engineers with great effect as proof that they considered themselves superior to the working class.

It was convenient for Stalin in his struggle for absolute political power that one of his main critics, Nikolai Bukharin, was associated with the technocratic camp.[70] This association was not an anti-Soviet conspiracy, as Stalin maintained, but merely a congeniality of viewpoint as well as a bureaucratic connection. Bukharin and his like-minded Party colleague A. I. Rykov often defended engineers and praised their approach to industrial planning. Bukharin even repeated the phrase "the future belongs to the managing-engineers and the engineering-managers."[71] Furthermore, for a brief time Bukharin headed the Scientific-Technical Administration. Stalin thus had the opportunity to strike two blows: one against his main rival and another against what he regarded as the arrogant aspirations of engineers.

And strike he did, with tragic effects. Shortly before Palchinsky's arrest in April 1928 a group of engineers was charged with sabotaging coal mines near the town of Shakhty in the Northern Caucasus.[72] In May the Shakhty engineers were brought to trial; five were sentenced to death, six imprisoned for life, thirty-eight sentenced to prison for one to ten years, and four were acquitted. Then, over two years after Palchinsky's arrest, from November 25 to December 7, 1930, eight leading Soviet engineers were brought to what was called the Industrial Party Trial on charges of conspiring during the 1920s to overthrow the Soviet government.[73] Palchinsky, already executed in secrecy, was named the head of the conspiracy.

These events were only the beginning of a reign of terror among Soviet engineers, several thousand of whom were arrested. There were only about ten thousand engineers in the entire Soviet Union at the time. In the end, about 30 percent of Palchinsky's colleagues were arrested—most of them thrown into labor camps with little chance of survival. The lucky ones were placed in special research and development prisons and assigned tasks by the government. Aleksandr Solzhenitsyn's novel *The First Circle* describes one of these prison laboratories, created in the wake of the Industrial Party Trial.[74]

Palchinsky's wife, Nina Aleksandrovna, knew after her husband's death that she too was in peril, since family members of "enemies of the people" were often imprisoned. To make matters worse, her old friends in Moscow and Leningrad were afraid to fraternize with her lest they also fall under suspicion. On August 16, 1929, she wrote to Kropotkin's widow, "I have been left without any funds, and no one has given me any help, all shun me and fear me . . . And I have found out what friends are now. There are very few exceptions."[75] She sought anonymity by fleeing to the provinces of Russia, where she took a job at the lowest rank of the nursing profession. Well educated and a strong defender of women's rights before and after the Revolution, she found it difficult to adjust to her new life. One evening she went to a local movie theater, where a film about the Russian Revolution was being shown. To her horror, the film featured a depiction of her husband, Peter Palchinsky, as an enemy of the Revolution. A person in the audience who barely knew her shouted, "We have a Palchinsky among us!" and pointed at Nina

Aleksandrovna.[76] She was promptly arrested by the local police and disappeared into the camps.

The Posthumous Secret Police Report

What about Palchinsky's activities and opinions did the Soviet secret police find most offensive? What specific charges did they level against him? The answers to these questions are found in the secret police report on the Industrial Party from the INION library.[77]

In this report Palchinsky is described as the ringleader of the "engineer-wreckers" who were trying to restore capitalism in the Soviet Union. The centers of the anti-Soviet conspiracy were the Institute of the Surface and Depths of the Earth and the Miners' Club, both organized by Palchinsky. A major charge against Palchinsky and his organizations was that they insisted on publishing "detailed statistics" on the mining and petroleum industries, which could be used by anti-Soviet organizations. Thus, the Soviet authorities were taking the same attitude toward industrial and labor statistics that the tsarist government had when Palchinsky had investigated the "labor question" in the Don Basin at the turn of the century. The tsarist government punished him by exiling him to Siberia; the Soviet government chose arrest and execution.

Much of the secret police interrogation of the "engineer-wreckers" reads like a medieval chronicle that was written primarily for moral edification; it does not provide reliable facts, but it shows instead what myths the secret police wanted to create. The engineers are described as traitors who had been bought off by Western companies, such as Shell Oil and the Nobel Company, and were following orders from abroad on how to destroy Soviet industry. They were incredibly greedy and would do anything for pay. They would use any means—legal or illegal—to achieve their goals. They capitalized on internal disputes in the Party, such as those concerning Trotsky or Bukharin, to depict Soviet rule on the verge of collapse.

Their plans for economic development were aimed at sabotaging Soviet industry by slowing growth through the establishment of minimal goals.

No one can attest that every engineer arrested by the Soviet authorities during the Industrial Party Trial was completely guiltless, but the evidence points overwhelmingly to their innocence. For Palchinsky, the alleged ringleader, we now have access to his personal papers, drafts of his speeches, private letters to his relatives, and detailed accounts of his business and private meetings. Because he was a compulsive collector of personal records, his papers include train tickets, receipts, notes taken at professional meetings, and reactions to articles and books indicated by underlined passages and marginal notes. Nowhere in this mass of information do we find any indication that he worked against the interests of Soviet industrialization. On the contrary, we find abundant evidence that he aided it in every possible way.

By the middle 1920s Palchinsky had become an enthusiastic supporter of industrialization under a socialist, rather than a capitalist, economy. In 1926 he asked, "How can there be any comparison between the development, for example, of oil and coal regions under capitalism and under the present socialist order? . . . Only the nationalization of the surface and depths of the earth permits ideal choices for a planned, rationalized governmental economy."[78]

But it was precisely Palchinsky's enthusiasm about the new possibilities under socialism that led him to believe that engineers, not capitalist owners of factories and mines, must play a strong role in creating plans for economic development. The main goal that Palchinsky assigned to his Institute of the Surface and Depths of the Earth was "to help the economic development of Russia by studying its natural resources and working out their most rational use."[79] In similar fashion, the function of the Miners' Club, which he also founded, was to give the Soviet government assistance in evaluating mining projects to "obtain the maximum degree of objectivity."[80]

Palchinsky may, in retrospect, appear naive in his belief that the Soviet government would appreciate such advice, just as he had misjudged the tsarist government when he sent it evaluations of the Don Basin mines, but his intentions were to assist his country by increasing its industrial strength and the welfare of its people.

3
Early Soviet Industrialization

The life and death of Peter Palchinsky provide vivid clues to the failures of Soviet industrialization policies. Palchinsky serves well as an example of the type of engineer destroyed in early Soviet history who was sorely needed in later years. His fate and that of like-minded engineers of the late 1920s had a negative effect on industry in the Soviet Union for many years—just as the similar elimination of enterprising farmers, the so-called kulaks, damaged agriculture. As Gorbachev, Yeltsin, and their colleagues strove to revivify agriculture and industry in the late eighties and early nineties, they had to overcome the lingering effects of the removal of the economy's best representatives and their replacement by individuals determined, at all costs, to avoid their predecessors' fate.

From the very beginning of the First Five-Year Plan in 1927, before Palchinsky's death, the engineering principles that he and his colleagues promoted were consistently ignored. They had maintained that industry in a socialist society must rank workers and local populations as the highest priority; for without healthy, educated, and well-motivated workers industrial expansion would be illusory and ill-fated. They also emphasized that industrial expansion must be conducted according to rational principles and planned for the long term. The Five-Year Plans, so celebrated in the Soviet Union and abroad, not only ignored these principles but betrayed the enthusiasm of workers who enrolled in the building of socialism.

The Soviet industrialization drive in the late twenties and early thirties coincided with capitalism's greatest crisis. While America and other industrialized countries struggled through the Great Depression, socialist Russia developed at a frenetic pace. Western correspondents in Moscow sent home stories of the heroic achievements of Soviet miners, steel mill workers, lathe operators, and builders of ship canals and hydroelectric dams. Coming at a moment when the unemployed lined up in the soup kitchens of Detroit, Manchester, and Pittsburgh, for many observers in the West the Soviet industrialization effort became a vision of the socialist future of all humankind. Even today the memory of these events remains strong outside the former Soviet Union. And still today the technical and social wisdom of these projects has been largely unexamined. Although this book cannot attempt a thorough reevaluation, enough is now known to see that Soviet industrialization projects were badly flawed from an engineering standpoint, flagrantly wasteful of the faith of those Soviet workers who supported them, and dreadfully costly of the lives of those who worked on them voluntarily and involuntarily.

The engineering flaws were not unknown at the time. Indeed, engineers with pre-Revolutionary training, including Palchinsky, enumerated most of the problems before the projects began and provided ammunition to critics who were inclined to describe them as saboteurs or "wreckers." The evidence indicates, however, that most engineers from the old regime were enthusiastic about the potential offered by a planned socialist economy, and spoke out only against irrational choices by the Stalinist leadership.

Three of the monumental projects in the early Five-Year Plans were the building of the world's largest hydroelectric plant on the Dnieper River (Dneprostroi), the construction of the world's largest steel plant along with the west Siberian city of Magnitogorsk (Magnitostroi), and the digging, in record time, of the White Sea Canal linking the Baltic and White Seas (Belomorstroi). The old engineers participated in all three of these projects. Peter Palchinsky worked on two of them, Dneprostroi and Magnitostroi, and commented on the third, Belomorstroi (the suffix -*stroi* comes from *stroika*, construction project). His engineering colleagues imprisoned after his death were the main advisors to Belomorstroi. It is enlightening to look briefly at the role that technical advice played (or failed to play) in each of these projects.

The Great Dnieper Dam (Dneprostroi)

The hydroelectric station near Zaporozhye on the Dnieper River, centered on the impressive Lenin Dam, was one of the most fabled projects of the First Five-Year Plan. It was the forerunner of the other great construction projects of early Soviet industrialization—among them the steel city of Magnitogorsk, the coal city of Kuznetsk, and the White Sea Canal.

Of all these gigantic projects the hydroelectric plant on the Dnieper was probably the most sensible from an engineering standpoint. It was preceded by more careful analysis than most of the others, it relied less on prison labor, it utilized more labor-saving machinery, and its Soviet administrators were more willing to listen to expertise—both foreign and domestic—than was the case in most later projects. Colonel Hugh Cooper, the maverick American engineer (whom Palchinsky knew), was a major consultant to the project, as were several German engineers.[1] A number of foreign companies were involved. Cooper and other specialists insisted on the use of giant cranes and other modern equipment, brought in from abroad. Dneprostroi also benefited greatly from the energy of thousands of Soviet workers who genuinely wished to help build a new socialist society. As the first of the great projects, the Dnieper dam project also allowed more room for dissenting advice than the later ones. In the early and mid-twenties, when the Dnieper dam was planned, an engineer still could raise technical questions without immediately being labeled a "wrecker," or a foreign engineer could propose changes in construction plans without being accused of serving the interests of foreign capital.

The more closely one examines the Dnieper dam project, however, the more one realizes that it, too, suffered from all the defects of the later ones, even if in less stark forms. The decision to build the dam was flawed economically, and was even more dubious socially and ethically. Dneprostroi established patterns of the abuse of labor and local populations, which became more evident and flagrant in later projects and contributed to the decades-long disillusionment of Soviet workers. It was just the first among many experiences that gradually reduced support for the Soviet government among workers and peasants.

Had all the costs, social and economic, of the Dnieper dam been considered more carefully, and had the benefits of a single enormous hydroelectric power plant been weighed against those of several small ones, including thermal power plants, a different decision probably would have been made. These alternatives, now quite obviously more desirable, were outlined by Russian engineers during the early planning stages. The final decision to go ahead with the giant dam was based not on technical and social analysis but on ideological and political pressure. Stalin and the top leaders of the Communist Party wanted the largest power plant ever built in order to impress the world and the Soviet population with their success and that of the coming Communist social order. As Anne Rassweiler, a historian of the project, observed, "Its economic rationality was never proven . . . Clearly the decision to build Dneprostroi was made on other grounds."[2]

Many engineers, including Peter Palchinsky, warned against a hasty decision to go ahead with the big dam. Perhaps the most outspoken was R. E. Klasson, a specialist in electric power. He pointed out the plentiful coal supplies near the proposed dam, and observed that the decision to build a hydroelectric station or a thermal one should be a matter of calculating social and economic costs. He maintained that a thermal plant would be needed in any event, since the level of water in the Dnieper River was inadequate to generate electricity from December to February. Furthermore, the normal flow of the water was so slow that turbines would have to be large and therefore expensive—in fact, the largest ever built.[3] In addition, occasional dry years would force reliance on thermal electric plants even in summer months. Klasson's recommendation was to start with the construction of one or two steam plants and to expand in step with the demand for power in the region, combining hydroelectric and thermal plants as needed.[4]

Palchinsky admonished the government not to plan enormous hydroelectric plants, such as Dneprostroi, without regard to the distance between where the energy would be used to where it would be generated. The likely consequences, he predicted, would be huge transmission costs and declines in efficiency. He also was dismayed that the plans for Dneprostroi were going ahead before complete geological, hydrological, or topographical maps had been prepared. Furthermore, no good studies had been made of flow patterns of surface and underground water in the area. As a result, no one knew just how large an area the 35 meter (96 feet) high dam would flood.

To clear the area for the reservoir, over ten thousand villagers were forced out of their homes. Most of them were German Mennonites, prosperous and industrious farmers. The loss of their farmlands was not included in estimates of the expense of Dneprostroi. Many years later a prominent Russian hydrologist would calculate that the hay annually harvested from these lands would, if burned as fuel, have produced as much energy as was generated by the hydroelectric power plant.[5] Even allowing for some latter-day exaggeration here, the loss of these lands was a great hidden cost.

Aside from the economic expense of taking this farmland out of production was the human cost to the farmers themselves, which none of the Soviet planners considered. A strongly religious people with above-average wealth in land and livestock, the Mennonites were seen as ideological opponents to the Soviet order. Even before their lands were flooded, their buildings were taken over as workers' barracks, and they were offered jobs in the workforce at the dam. If they agreed to this change in their status, they were considered voluntary workers, along with the majority of the almost forty thousand workers who came to the site. Some of the Mennonites made the transition well, accepting what they saw as God's fate. Others resisted, were arrested, and became a part of the forced labor contingent, never as large at Dneprostroi as at most of the later great construction sites. These prison laborers were marched to work under armed guard and were given the most difficult jobs.

The Lenin Dam was built on a flood-plain river, unlike most hydroelectric power plants in other countries, and a large number of people and a great area of land were affected by the resulting reservoir. Later Soviet hydroelectric power projects continued this practice on an even larger scale. The building of the Rybinsk hydroelectric power plant required the relocation of 497 villages and 7 cities. One writer has estimated that 120,000 square kilometers, equivalent in area to four Belgiums, were eventually flooded by the reservoirs of such power plants.[6]

Initially the builders of Dneprostroi promised to provide adequate housing and cultural facilities for the workers. As the project proceeded, falling behind in its schedule and grossly exceeding the estimated costs, the workers' needs were more and more neglected. By the time the project was at full speed, the workers were living and laboring under miserable conditions: "Barrack dwellers complained of snow drifting through rooms. Tent dwellers endured temperatures below −13 degrees C in the winter, and tornado-strength winds whipped tents away in the summer of 1929. Crowding, dark, and noise were endemic. Toilet facilities were inadequate and frozen in the winter months."[7] In the following months, conditions became worse and food grew increasingly scarce. Flour had to be delivered to the bakeries at night under armed guard to prevent theft. The

poor nutrition contributed to outbreaks of disease. Tuberculosis, ty-
phus, typhoid fever, and smallpox spread throughout the barracks
and tents. No one knows to this day how many people died.

Despite these hardships, construction was completed and the
great dam was put into operation. It became a symbol of the socialist
order, and was a centerpiece of the Soviet exhibit at the World's Fair
in Chicago in 1939. Destroyed and rebuilt twice during World War
II and expanded several times since, it is still in operation today as
one of six hydroelectric dams on the Dnieper River.

It remains unknown how efficient or useful the Dnieper hydroelectric power plant is. Just as adequate studies were not made before
its construction, no complete evaluation has been made since. Any
such evaluation would include the full range of the project's economic, social, and ecological impacts. A Soviet environmentalist
observed in 1980: "The money spent to control erosion on the shores
of the Dnieper reservoirs and to combat the seemingly indestructible
blue-green algae has long since exceeded the short-term advantages
the power plant once yielded."[8] The original planners of the dam
should perhaps not be blamed for disregarding ecological effects,
since little was known in their time about environmental damage
from reservoirs. But they did knowingly ignore the calls of Palchinsky
and his fellow critics of the dam for continuing and complete studies
of the social and economic effects of such gigantic projects.

The Steel City of Magnitogorsk (Magnitostroi)

In 1929 construction began on a gigantic complex of blast furnaces,
open hearth furnaces, and finishing mills that would eventually produce each year almost as much steel as all of Great Britain. The
complex was built near the site of one of the country's richest iron
deposits, known as Magnetic Mountain because of the disturbances
it caused to the compasses of early explorers of the region. Actually
a series of five hills, the Magnetic Mountain region was remarkable

not only for the richness of the iron found there, but also for the ore's accessibility. It seemed reasonable to build Soviet Russia's greatest steel mill next to this geological wonder.

But was this really the best location for such a large steel mill? In articles published in 1926 and 1927, Peter Palchinsky complained that the Soviet government was going ahead with plans for the construction of enormous mining and refining operations in West Siberia, the Urals, and Ukraine without adequate studies of the geological resources, availability of labor, economics of transportation, and difficulty of supplying proper housing for the workforce. He noted that although everyone marveled over the rich ore of the Magnetic Mountain, no one had yet made a thorough study of the amount of iron it contained.[9] It was quite possible that after a few decades the ore would be exhausted, and yet the continuing presence of the world's largest steel mill would require the costly hauling of ore from other regions.

He noted that no coal was available near the projected city of Magnitogorsk, so that from the very beginning fuel for the voracious blast furnaces would have to be hauled in by railway. He also observed that the region was not served by waterways, although water is by far the least expensive means of hauling such heavy loads as iron ore and coal. In other countries, such as the United States, he pointedly observed, the steel mills are located not near the rich ore deposits of the Mesabi Range in Minnesota or the Marquette Range in Michigan, but hundreds of miles away in Detroit, Gary, Cleveland, and Pittsburgh—all cities with large labor forces, the first three connected to the sources of ore by water, and the last located near enormous coal deposits. The costs of building the city of Magnitogorsk and its mills might be so great, Palchinsky continued, that it might be wiser to expand steel production near less rich deposits of iron ore but at locations with better labor and transportation resources.

Palchinsky was quite enthusiastic about the idea of building a large steel complex, however, and he did not rule out the possibility, despite his doubts, that Magnitogorsk was the right place to do it. But he argued that the decision to proceed with construction must wait for the completion of thorough planning studies, which he thought could be carried out rather quickly. The selection of a site for an industrial facility must be based on multiple factors, no one of which—such as the location of the raw material—is governing. He called for the drawing up of gravimetric charts, magnetrometric measurements, and economic calculations, and for the use of the new methods of transportation and freight engineering. And, once again, he reminded the Soviet government that the most important factor of all was human beings. Adequate workers' housing and urban amenities were preconditions for such a gigantic complex.

Palchinsky's worries about the location of Magnitogorsk were ignored. The Soviet government announced that the new steel complex would be equipped with the latest technology and would surpass all Western competitors in size and quality. The last word in steel-making at the time was the United States Steel works in Gary, Indiana, and the Magnitogorsk mills were to be bigger and better. To fulfill this goal, the government brought American engineers, some of them from Gary, to help plan the mills. The foreign engineers were not asked whether Magnitogorsk was the right spot to build such a mill; they were simply told to advise on its construction. The American engineers were housed near the planned mill in a special settlement called "Amerikanka," where they enjoyed luxuries not available to their Soviet counterparts, such as private houses with modern conveniences and even tennis courts.

The Soviet government promised that the workers would soon have similar perquisites, that the proposed "garden city" of Magnitogorsk would provide the finest workers' housing in the world. The German architect and city planner Ernst May was brought, fresh from his innovative work in Frankfurt, to plan the new city.[10] May was an especially strong proponent of protecting workers from the deleterious effects of industrial pollution by separating industry from residences with greenbelts. Until his plans could be realized, though, the workers were housed in tents and barracks.

The garden city was never built. Pressure from Moscow bosses to meet strenuous work schedules and production quotas kept the building of workers' housing last on the list of priorities. The two hundred thousand workers who were brought to the city continued to live mostly in barracks, tents, and mud huts in conditions of filth and deprivation, without indoor plumbing and surrounded by open sewers. Ernst May's attempt to build his "socialist city" apart from the industrial pollution was constrained by the bureaucracy and eventually attacked in the press. The temporary workers' housing, which soon became permanent, was directly in the path of the fumes from the blast furnaces. In 1934 May fled the USSR in disillusionment, complaining that in Germany, under Hitler's rule, and in the Soviet Union, under Stalin's grip, "humanity had entered a period of decline similar to the medieval ages."[11]

Much worse than the regular workers' lives at Magnitogorsk were those of the thirty thousand or so kulaks, peasants dispossessed by the collectivization of agriculture and brought as prisoners hundreds of miles to work on the construction of the steel city. Laboring under armed guard, they did the heaviest and most disagreeable work. Without adequate food or clothing, housed in tents, an estimated 10 percent of them died during the first winter.[12] Later they were placed in barracks that housed forty to fifty families each. These barracks remained the domiciles of the "former kulaks" and their children for decades; only in the late 1960s and early 1970s were the survivors able to move to apartment buildings.[13] Even as late as 1989, 20 percent of the apartments in Magnitogorsk were of the communal type, in which each bedroom was occupied by a separate family, and the kitchen and bathroom were shared by all.[14]

After Gorbachev launched his reforms in the Soviet Union in the mid-1980s, residents of Magnitogorsk began for the first time to explore their own history. In 1988 a former kulak who as a child had been brought to Magnitogorsk as a prisoner described the experience:[15]

> As many as forty families were squeezed into a single freight car with barred windows. It was possible only to sit, not to lie down. For the necessities of nature there was a wooden bucket. For three days, while the train was under way, there was a heat wave. In the wagon it was stuffy. For a day and a half the door was not even opened . . . Mothers had children die in their arms.
>
> When we got to Magnitogorsk there was a cart by the tracks. We guessed—it wasn't for the living. From only the wagon in which we traveled, four little corpses were removed. More were carried out from other wagons . . . They put us up in canvas tents, six families to a large one; two to a small one. Each tent was numbered. The first months we all lived in that tent, and, of course, water penetrated the canvas. The ground underneath was frozen. People wrapped themselves in fur coats, animal skins, rags, whatever was brought from the villages.

Such stories did not fit with the official mythology of the building of the great steel mills of Magnitogorsk, which portrayed it as the feat of thousands of "enthusiasts," volunteers who came to the site to help construct socialism. In the late eighties, the newly free press in Magnitogorsk debated "Who built Magnitka? Prisoners or enthusiasts?"[16] Paradoxically and sadly, the answer to the question was both. One of the characteristics of industrialization under Stalin was the coexistence of volunteer and forced labor, of heroic self-sacrifice and violent coercion. It was a temporary and unstable dichotomy that could occur only in a society that was simultaneously undergoing a social revolution with substantial support from below while being enslaved from above.

One of the enthusiasts who came to help build Magnitogorsk was a young American, John Scott, who left a remarkable account of his five years there.[17] The son of two well-known American radicals, Scott Nearing and Nellie Seeds, John Scott went to the Soviet Union at the height of the Depression seeking an alternative to failing capitalism. Although he was appalled by the suffering and cruelty that he found in Magnitogorsk, Scott also was a witness to the dedication and heroism of many of the workers. He wrote of Magnitogorsk after his return, "The people were studying, looking forward, striving to build something in which at least many of them believed."[18]

Soviet ideology justified the coexistence of prison and voluntary labor by portraying the kulaks as enemies of socialism deserving of their fates while celebrating the volunteer workers as the leaders of the new order. But eventually the coercion, deprivation, and injustice that were endemic in the new regime destroyed the beliefs and confidence of even its supporters. John Scott became a staunch critic of the Soviet Union after he returned to the United States. The workers of Magnitogorsk eventually lapsed into the apathy characteristic of most of the Soviet labor force.

Palchinsky voiced the suspicion in 1926 and 1927 that the inadequately explored iron deposits of the Magnetic Mountain would be more limited than expected and that the location of the giant steel mills away from waterways might turn out to be unwise. By the early 1970s that skepticism proved correct; the exhaustion of local iron ore required the mills to import ore by train from other regions, as they had from the very beginning the coal for the blast furnaces. From this time on, Magnitogorsk was forced to obtain both of its heavy raw materials from a great distance over land, rather than by cheaper water transport. Magnitogorsk became a monument to inefficiency.

In 1987 the young historian Stephen Kotkin was the first American to live in Magnitogorsk for any length of time since John Scott had resided there in the 1930s. He found a dirty and dispirited city surrounding hopelessly obsolescent steel mills. Far from the garden city of socialism that Scott and many Soviet citizens had expected would be created, Kotkin observed "entrenched alcoholism, recurrent shortages of consumer goods, a severe housing crisis extending well into the future, a ubiquitous black market, a crumbling or nonexistent urban infrastructure, almost unfathomable pollution, and a health catastrophe impossible to exaggerate."[19]

The White Sea Canal (Belomorstroi)

The building of the White Sea Canal, another of the large projects of the First Five-Year Plan, was a nightmare. It not only ignored the engineering principles of Palchinsky and his colleagues but it was also an obscene violation of human rights. In the building of the Dnieper dam, forced labor played only a minor role. In the great project of the steel city of Magnitogorsk, many more prisoners worked alongside free laborers, but they still provided only part of the workforce. At the construction of the White Sea Canal almost all the workers, from the supervising engineers down to the lowliest laborers, were prisoners, detained for ideological reasons rather than convicted of actual crimes. They toiled under conditions of unimaginable cruelty; approximately two hundred thousand of them died during less than two years of construction, an average of ten thousand a month (the rate was much higher in the Arctic winter and lower in the summer).

The goal of the White Sea Canal was to connect the Baltic and White Seas, a dream that went back to the time of Peter the Great. Ships traveling between Arkhangelsk and St. Petersburg were forced to go all the way around Scandinavia in the dangerous North Atlantic and through the Baltic Sea, a voyage requiring many days, or even weeks. Tsar Paul I at the beginning of the nineteenth century commissioned a feasibility study for the canal, but abandoned the idea in the face of obstacles presented by the rocky terrain of Karelia, through which the canal would have to pass. Stalin raised the issue again in the 1920s, and spoke of the advantages that a canal suitable for oceangoing vessels would bring to Soviet industry in the Leningrad and Karelian region. Furthermore, he noted, naval vessels could be easily transferred back and forth between the Baltic and White Sea fleets.

Canals, like hydroelectric power stations, were suitable symbols for the building of socialism. The White Sea Canal became the subject of documentary books (scandalously falsified) and novels. A popular brand of cigarettes was even named after it. The accounts of the canal's construction published in the Soviet Union totally ignored the human costs of the project. An English-language version of one of these books, which glorified the building of the canal and concealed its cruelty, was described by the duped British author who wrote the introduction as "diverting" and "exciting."[20]

Stalin was a great admirer of canal projects, and he was fascinated by the role of engineers in their construction, especially engineers whose expertise was necessary but who could not be trusted because of their political views. Two of his favorite novels before the Revolution were Aleksandr Bogdanov's *Red Star* and *Engineer Menni*.[21] In these works of science fiction, the builders of socialism on the planet Mars have to rely on an engineer named Menni, educated before the Socialist Revolution, who is both brilliant and traitorous.[22] Menni recommends a path for a canal that purposefully delays construction and causes the deaths of many laborers. Menni is arrested, the mistakes are rectified, and the canal is completed. Stalin believed that, if kept under surveillance, even hostile technical specialists could be forced to yield their expertise for the benefit of the state. Yet the final results of the White Sea Canal project provided justification for Peter Palchinsky's views, expressed in a 1921 essay, that the coercion of engineers and workers would produce flawed, even "monstrous" results.[23]

The engineers who led the project—N. I. Khrustalev, O. V.

Viazemskii, A. G. Anan'ev, V. N. Maslov, K. A. Verzhbitskii, and K. M. Zubrik—were not permitted to question the wisdom of the project itself. That the canal would be frozen half the year and that a modernization of the existing railroad, usable year-round, might be more sensible, were not included in the analysis. As prisoners, the engineers were allowed to submit suggestions to their police supervisors only for the path of the canal and the method of construction. They came up with two choices for its location. The "western variant" would go through places named Vodorazdel and Segozero, and join up with the rivers Segezha and Kumsa. The "eastern variant" would traverse Lake Vyg and a deepened Vyg River. The advantages of the western option were that it would be deep and would have a secure supply of water. Its disadvantages were that it would take longer to build and would require more dams and more mechanized equipment.

The engineers recommended the more reliable and expensive western option. They feared that the water supply for the eastern variant, which was largely dependent on snow runoff in the spring, would not be adequate, especially in years of light snowfall. The deeper western path could use large concrete dams to maintain the water supply even in dry years. In reply the engineers were told that the canal absolutely had to be completed in the extremely short time of twenty months, and that no mechanized equipment or concrete could be used, since they would require expenditure of foreign exchange. All building materials must be available on the spot, which meant construction was limited to wood, dirt, and stone. All work was to be accomplished by human beings and horses—no excavators or mechanical cranes of the sort that had been used at Dneprostroi at the insistence of Hugh Cooper.

The political vulnerabilities of the imprisoned "bourgeois specialists" made their position impossible: if these engineers continued to defend the western path for the canal, they knew they would be accused of trying to sabotage the project, like Engineer Menni, or of having economic loyalties to the capitalist west from which equipment would be purchased. Although they were already prisoners, they could be further punished—by reduction to the position of the hardest manual labor, transfer to a prison elsewhere, or even execution. Against their better judgment, they assented to the eastern plan, and threw themselves into the supervision of hundreds of thousands of prisoners using the most primitive means of construction.

The jargon of the camp sarcastically described the work methods.[24] A "Belomor Ford" was a wooden platform on four wooden rollers pulled either by a pair of horses or by dozens of prisoners. A "hydraulic derrick" was a wooden crane powered by horses. An "excavator" was a crew of workers, many of them women, armed with wheelbarrows. Large stones were moved by placing rope nets around them and having horses or prisoners haul on the lines. Trees were pulled down in the same way.

The engineers were told to construct the canal walls with a minimum of concrete or metal, so they used timber cribs filled with stones and dirt. In the winter, ice would often end up in the cribs; when it melted in the spring, the cribs had to be reconstructed. The engineers requested metal for the lock-gates but were again refused. Engineer Maslov designed wooden gates that would work for only a few years before they rotted.[25]

The prisoners lived in tents or wooden barracks, sometimes even in the open. Their food was inadequate for survival. Many years later a few survivors wrote about the life (and death) of the forced laborers. D. P. Vitkovskii described the winter work:[26] "At the end of the workday there were corpses left on the work site. The snow powdered their faces. One of them was hunched over beneath an overturned wheelbarrow, he had hidden his hands in his sleeves and frozen to death in that position. Someone had frozen with his head bent down between his knees. Two were frozen back to back leaning against each other . . . At night the sledges went out and collected them."

On May 1, 1933, the head of the secret police Genrikh Yagoda reported to Stalin that the canal had been completed on time. In July of the same year Stalin and several of his top assistants took a cruise on the canal on a small steamboat.

From the beginning the White Sea Canal failed to live up to its specifications. It was so shallow, even in years of the deepest snowfall, that oceangoing vessels could not use it. The transfer of large naval vessels, one of the justifications for the canal, was impossible. Within a few years the walls of the canal began to crumble and the lock-gates to collapse. After World War II the rebuilding of the entire canal began. In many places the new canal ran parallel to the first and so had the same water shortage problems, but at least it had better walls and metal gates on its locks. In 1966 Aleksandr Solzhenitsyn spent a summer day at the canal; during eight hours only two barges passed, loaded with timber and going in opposite directions.[27] The locks guards admitted that there was little canal traffic. But the Belomor remained a part of the folklore of Soviet industrialization.

Technocracy, Soviet Style

After 1930 engineers in the Soviet Union turned away from the broad social and economic questions that Palchinsky believed were intrinsic to the engineering task. One reason for this change, especially during the thirties and forties, was fear. Following the purge of the early thirties, Soviet engineers well understood that if they wanted to stay out of trouble they must concentrate on the narrow technical tasks assigned to them by the Party leaders. An American engineer who visited his Russian colleagues in the Soviet Union shortly after the Industrial Party Trial in 1930 reported, "from the very first, I received the distinct impression that there was fear in the very air the Russian engineer breathed, and if he did not dare practice his profession with integrity, it was because he was judged politically before he was accepted professionally."[1]

This pervasive fear caused engineers to avoid controversy. They stopped raising questions about workers' safety or workers' housing because they knew that such questions could irritate their bosses — directors who were concerned primarily with meeting the production quotas set for their factories and mines. But even sticking closely to their assigned technical tasks did not guarantee that engineers would avoid political trouble, since their ability to increase output was constantly being judged. A failure to meet quotas could become a "political" mistake in the eyes of the leaders of the local Communist Party organizations. Therefore, many engineers tried to get out of production completely, fleeing to research and development institutes where their possible failures might be less easy to identify. An American historian who noticed this phenomenon called it "the flight from production."[2]

The second reason for the engineers' narrowing definition of their work, with long-term effects down to the present day, was a change in their education. The training of engineers was taken away from the Ministry of Education, which had been interested in broad education, and turned over to the industrial ministries, whose institutes set restricted and instrumental goals for their graduates.[3] Professors at these engineering institutions avoided issues touching on politics and social justice, and concentrated on science and technology. The Soviet Union's engineering institutions began to produce the new breed in great numbers, and the newcomers rather quickly replaced the engineers with pre-Revolutionary educations. In the decades after 1930, the Soviet Union trained more engineers than any other country in the world; yet these engineers, with a strikingly circumscribed vision, aimed only at increasing production, to the neglect of all other factors. The education received by the new Soviet engineers was not only more restricted than that of their predecessors of the tsarist period, but also narrower than that of their colleagues in other countries.

One of the ways in which the new Soviet engineers differed from engineers elsewhere first became apparent to me in 1960 when I came to Moscow as an exchange student at Moscow University. Five years earlier I had received a degree in chemical engineering from Purdue University. At Purdue I had been distressed by the narrowness of the curriculum. The few elective courses I had time to take were inadequate windows on the large and complex world beyond thermodynamics and differential equations that I wanted to explore. After graduation I worked briefly as an engineer and then returned to graduate school, this time to Columbia University, to study the history of science and technology, particularly in Russia. The year in Moscow was for dissertation research.

In some ways I still thought of myself as an engineer and was therefore interested in finding Soviet engineering students with whom I could exchange views. There were no engineering students at Moscow University or at Leningrad University, which I also visited; all engineers, I was told, were trained at special technical institutes. Finally, during a student excursion outside Moscow I met a young woman who said that she was an engineer. "What kind of an engineer?" I asked. "A ball-bearing engineer for paper mills" was the reply. I responded, "Oh, you must be a mechanical engineer." She rejoined, "No, I am a ball-bearing engineer for paper mills." Incredulous, I countered, "Surely you do not have a degree in 'ball-bearings for paper mills.'" She assured me that she did indeed have such a degree.

In succeeding decades I became fascinated by the history of the Soviet engineering profession and the unusual influence engineers had on Soviet political and economic development. After I began research on Peter Palchinsky I learned that most of the distinctive features of engineering in the Soviet Union can be traced back to the time of his death, the end of the 1920s and the beginning of the 1930s.

Peter Palchinsky's views on engineering education starkly contrast with those of the engineering establishment in the half-century under Stalin, Khrushchev, and Brezhnev. During this time, engineering dominated all other modes of education—mainly because no Soviet educational institutions offered nonspecialized education in the liberal arts, or general studies of the type so common in the United States, Britain, and other countries in the West.[4] Soviet graduates came either from universities, of which there were 40 in 1959, or from specialized institutes, of which there were 656 (not counting correspondence and evening schools).[5] Neither offered educations that would have been considered broad in the West, but the best universities, such as Moscow and Leningrad, were certainly superior in this regard to the institutes. Unfortunately, the universities' graduates were swamped by the institutes'; between 1930 and 1960, 88 percent of graduates of higher education came from specialized institutes outside the university system.[6]

The humanities, as known in the West, played almost no role in Soviet education in the Stalinist and post-Stalinist periods. Instead, Soviet students at an early point chose a specialization that entailed training for a certain occupational goal. And these specializations were quite narrowly defined; as Nicholas DeWitt, the leading authority on Soviet education in the 1960s, wrote: "Professional specialization in the Soviet Union is much more pronounced than anywhere else in the world."[7] It was this intense specialization that produced my Moscow acquaintance, the ball-bearings-for-paper-mills engineer.

Some specializations were narrower than others. A Soviet student who majored in literature, foreign affairs, or art history obviously gained some knowledge of social issues. But in the engineering institutes there was no tradition of general or liberal education of the sort that even dedicated engineering schools like Purdue, Caltech, or MIT recognized. Students in Soviet engineering institutes did not major in mechanical engineering, civil engineering, or electrical engineering, as was the practice in most other industrialized countries, but instead in one of hundreds of subspecialties. Harley Balzer of Georgetown University has described this trend as a product of the thirties: "Each commissariat sought to train its own staff in specialties so limited that they bordered on the absurd . . . The Commissariat of Light Industry included engineering specialties for the compressors in each type of machinery. The Commissariat of Heavy Industry insisted on separate engineers for oil-based paints and non-oil based paints. The Commissariat of Agriculture trained agronomists for individual crops and veterinarians for each type of animal."[8] Speaking of the situation thirty years later, Nicholas DeWitt observed:[9]

The very number of specialties in Soviet engineering education reflects the continual narrowing and redivision of traditional engineering training. Thus mechanical engineering is broken down into several dozen related but perilously narrow specialties such as mechanical engineering in agricultural machinery, in machine tools, in casting equipment, in automobiles, in tractors and in aircraft engines. There are training programs for metallurgical engineering specialists in copper and alloys, specialists in the drilling of petroleum and gas wells or in the exploration of coal deposits; for civil engineering specialists in bridge design, in building large-scale hydrotechnical structures or in erecting industrial buildings. This fragmentation is characteristic of every field of engineering.

There was even a specialty in gyroscopic instruments, devices, and systems.

Every student in every Soviet higher educational institution was required to take courses in Marxism. In engineering institutes in the fifties and sixties there were three such courses, the only ones outside the technical curriculum: history of the Communist Party, dialectical materialism, and political economy. The textbooks used in these courses were designed for ideological indoctrination rather than for scholarly inquiry. The textbook on political economy is particularly revealing. Palchinsky had suggested that all engineers be required to take a course in political economy, but what he had in mind was the study of the complex interactions of society, economics, and industry, with a presentation of the ideas of major economic theorists. What Soviet students took instead was an overview of the Marxist stages of history. The three main sections of the Soviet textbook in political economy in the post-Stalin Soviet Union are "Pre-Capitalist Modes of Production," "Capitalist Mode of Production," and "Socialist Mode of Production." The book constantly reiterates the advantages of the Soviet economy: it is based on the public ownership of the means of production, it gives first priority to heavy industry, and this industry is concentrated in enormous factories containing thousands of workers. No concession is made to the opinion of Palchinsky and many others that a mature economy would necessarily be heterogeneous, combining large and small industries, private and state enterprises, individual and corporate management.

The Soviet textbook on political economy of 1958 contains 231 footnotes, but not a single one refers to a non-Communist source.[10] Over a third of the references are to Marx, Engels, or Lenin. The rest are to resolutions of the Communist Party, works of Stalin, Khrushchev, and Mao Tse-tung, and Soviet government laws and resolutions. (In later editions the references to Stalin and Khrushchev are replaced by references to Brezhnev, and the citations of Mao have been deleted.) There is no recognition of the host of economic theories developed in nonsocialist countries, no discussion of industrial management, no presentation of business economics. Is it any wonder that after the collapse of Communism at the end of the eighties Soviet engineers and industrial managers had such difficulty in adjusting to a market economy? They did not even possess the basic vocabulary necessary for understanding it.

The textbooks for the other two nontechnical courses that Soviet engineering students took during their four to six years of study were even less informative about the complex world around them than the one on political economy. The text on dialectical materialism contains a sterile treatise on the "laws of the dialectic" in nature. It soon became infamous among Soviet students as the most boring of their obligatory studies. The textbook on the history of the Communist Party of the Soviet Union presents a severely distorted account of Russian and Soviet history, always depicting the Communist Party as the "vanguard of the proletariat" and the director of the country's destiny. Such past leaders as

Trotsky and Bukharin appear only as cardboard figures, the heads, respectively, of the "Menshevist-Trotskyite Opposition" and "Bukharinite Anti-Party Group of Right Opportunists."

In sum, engineering students in the Soviet Union received a stunted and narrow education; it was intellectually impoverished, politically tendentious, socially unaware, and ethically lame. It would have been bad enough if the students who received it had stayed in the factories and research and development centers of the Soviet Union. But most of the leading political figures of the latter-day Soviet Union shared this educational background. In their new positions of power these ill-informed technocrats helped to determine the very mode of life of their fellow citizens.

The new engineers, woefully ignorant of the social and economic issues that Palchinsky considered so important, not only came to dominate Soviet industry but also provided the new generation of Communist Party leaders, replacing the Old Bolsheviks as they died off. Indeed, by the 1960s and 1970s so many of the top Party and government functionaries had engineering backgrounds that American specialists on the Soviet Union began to comment that an engineering education prepared people for high political offices in the Soviet Union as a legal education did for political leaders in the United States.[11]

Leonid Brezhnev, ruler of the Soviet Union for seventeen socially repressive years, received through evening study at the M. I. Arsenichev Metallurgical Institute an engineering degree in the production methods of rolled steel. He was not unique. Engineers were represented among the Soviet Union's political elite in proportions unparalleled among the other industrialized countries of the world. Between 1956 and 1986 the percentage of members of the Politburo, the top political body in the Soviet Union at that time, who had received their educations in technical areas rose from 59 percent to 89 percent.[12] If one defines technocracy as rule by people who were educated in technical subjects, the Soviet Union by the last quarter of this century was clearly a technocracy. And this was a technocracy ruled by engineers with more narrow educations than anywhere else in the world. The poet Boris Pasternak protested, "Does a canal justify human sacrifices? He is godless, your engineer, and what power he acquired."[13]

The restricted engineering educations of the great majority of the Soviet Union's top administrators influenced their management style and policy preferences. Trained to believe that the biggest enterprises were the best, they built on a grand scale. Even as late as 1992, after decentralization had begun, 75 percent of Russia's industrial enterprises had more than a thousand workers each.[14] These enterprises were inevitably flawed from the standpoint of investment of resources, environmental considerations, and social costs. Top administrators, most of whom were former engineers enamored of mammoth construction projects, knew precious little about economics and cost-benefit analysis, not to mention sociology and human psychology. Palchinsky would have considered them technicians, not genuine engineers.

Large-scale Soviet construction projects included not only the most ambitious hydroelectric power plants and canals of the twentieth century, but also the largest nuclear power installations ever built. As we learned after Chernobyl, those nuclear power plants were constructed with minimal safety features, and with little regard for emergency evacuation of the surrounding population.[15] Large as these projects are, even more breathtaking ones have been proposed; an example is the Northern Rivers Project, designed in the eighties, which would reverse the flow of several of Siberia's largest rivers in order to provide irrigation for Central Asian agriculture. Called the largest civil engineering project in history, it would have catastrophic environmental effects. Favored by land reclamation engineers and Central Asian political leaders, the Northern Rivers Project was vehemently opposed by environmentalists, by Russian nationalists who feared that flooding would destroy cultural monuments, and by several leading economists who questioned its cost-effectiveness. Soon after Gorbachev came to power, the Northern Rivers Project was shelved, although some Central Asian leaders still favor it.[16]

Housing construction in Soviet cities displayed the limited vision of engineers concerned more with ease of building than with aesthetics. The visitor to a contemporary Soviet city can witness row after row of nearly identical apartment blocks made of modular construction. The apartments inside are of only a few types. When the workers in Magnitogorsk were finally moved out of barracks into apartment blocks in the seventies, the variety of apartments was so limited that all the dwellers in the city knew the technical names of each. To know that your neighbors lived in "type A" or "type B" was to know everything about their abode, down to the location of the closets and the toilet. In some cities, such as Novosibirsk, quite a few Soviet citizens resisted moving from little log cabins to the new apartments, even though the cabins had no central heat and usually no running water. The cabins did have a touch of individuality, though, often displaying a bit of colored wooden fretwork around the windows that added a modicum of decoration to otherwise dreary lives. The resisters spoke of the "soulless" character of the monotonous new concrete apartment buildings.

In the fifties, sixties, and seventies, the same uniformity prevailed throughout the Soviet Union in clothing, appliances, street lights, park benches, and furniture. When I was a student in Moscow in the early sixties practically every apartment had identical overhead lampshades (abazhury), colored an iridescent orange. The aesthetics of an entire society were being dictated by the philistine tastes of the narrowly educated engineers who built the material objects and who ran the government. Only in the eighties, with the influx of foreign goods and the growing awareness of urban design, did the situation begin to change. One reason for the flourishing interest today in architectural restoration of pre-Revolutionary buildings is the yearning for a more variegated and sensually rich culture than the one fostered by the technocratic rulers of the country.

Soviet agricultural policymakers also sought a technological fix for an essentially economic and social problem. Their original preference for collectivized agriculture was based not only on the principle of socialist ownership of the land but also on a conviction that modern agricultural machinery, such as tractors and combines, could not be fully utilized as long as the land was divided into small private plots. There was a certain justification for this belief, as the average size of farms all over the world has grown during the last fifty years, but the Soviets' policy, determined by technology, was not sufficiently attuned to economic and psychological aspects that can make the difference between a hardworking private farmer and a listless state employee. As a socialist, Palchinsky would have probably approved of the common ownership of land, but he would have complained that Soviet agricultural policies neglected the most important component—the needs of the human being who is the producer.

When it became clear to Khrushchev in the fifties that Soviet agriculture was in deep trouble he reached once again for a technocratic solution: the extension of massive, mechanized state farms to previously uncultivated lands.[17] This program soon ran into trouble because it combined the problems of collectivized agriculture with the difficulties of raising crops on arid lands. Even after abandoning Khrushchev's utopian schemes for agriculture, the Soviet government under Brezhnev continued to promote large mechanized collective and state farms. In the end, however, the belief that such farms would prove their superiority as soon as sufficient machinery were available proved unwarranted. In the seventies the Soviet Union produced more tractors and combines than any other country in the world. Yet all this equipment could not solve the motivational problem that continued to cause low productivity in agriculture and poor distribution of its products.

Signs of Unrest among Soviet Engineers

The limited educations of Soviet engineers and of the political leaders who came from their ranks contributed to the narrow technocratic policies that governed the USSR during most of its existence. It would be a mistake, however, to think that attitudes among engineers remained frozen after the early thirties. By the fifties and sixties most Soviet engineers had overcome the fear that dominated the profession in the thirties and forties. Their support for the regime instead stemmed from involvement in the system . At the same time a small minority became increasingly aware of the damaging consequences of the silence that had been imposed on their profession.

Finding evidence of these rare criticisms of Soviet policies by engineers is more difficult than for most other professions. With a few exceptions, of whom Palchinsky was a prominent example, engineers everywhere do not often write essays of social and political criticism, and in the USSR the disincentives to do so were particularly great. Furthermore, during the forties and fifties most Soviet engineers had become thoroughly integrated into the military-industrial complex and knew that rocking the boat would almost surely harm their interests. Soviet engineers were, indeed, often among the most stalwart supporters of the economic and political order. The dissidents of the sixties and early seventies, writing critiques of the regime for publication in the underground press (samizdat), were typically natural scientists or members of the literary intelligentsia. Nonetheless, some signs of emerging political awareness among engineers can be found.

At a celebration of the end of World War II Stalin gave a toast to the low-echelon technical specialists who had helped bring about the defeat of Germany. He undiplomatically but accurately referred to these specialists as "screws in the great governmental machine." Some of the engineers and engineering students at the Central Aerodynamics Institute in Moscow responded by organizing a demonstration in which they lined up in a long column, packed as tightly as possible, and marched up and down the corridor of the institute, chanting, "We are screws, we are cogs in the machine." They were not punished, probably because they kept their demonstration within the walls of their institute and because they were among the most valuable engineers for the continuing Soviet military effort.[18]

A particularly important event for engineers was the appearance, in 1956, of Vladimir Dudintsev's novel *Not by Bread Alone*. The hero, Lopatkin, was an engineer who had devised a new method for casting sewer pipe. The story of his struggle to obtain the approval of the Soviet bureaucracy for his industrial innovation struck a sympathetic chord with many Soviet engineers who had faced similar problems.

Working engineers (those not holding full-time Party positions) in the post-Stalin period underwent a decline in status and prestige.[19] Their wages fell sometimes below those of skilled workers. It is no wonder that many of them tried to move out of industrial positions into other lines of work, including careers in the Communist Party apparatus, even though there they had no alternative but to accept Party discipline. Even the elite engineers engaged in research and development believed that they were losing social status, especially after they were expelled from the Academy of Sciences in the early 1960s. This issue prompted a few of the engineers to find their voices and begin to defend their professional interests. I. P. Bardin, a leading Soviet engineer who had once worked in the U.S. Steel plant in Gary, Indiana, sharply defended the presence of engineers in the Academy against the criticism of natural scientists such as the chemist N. N. Semenov.[20] Nonetheless, the engineers lost the battle, and the decline in their status continued. By the seventies and eighties, engineering institutions in the Soviet Union began to have difficulties filling their freshman classes. Yet the engineers were still strikingly quiescent.

In the seventies and eighties a few professional engineers voiced concerns about the inability of Soviet industry and Soviet military technology to keep up with their Western competitors. The engineers believed that the fault lay not with the engineers but with the bureaucratic obstacles they faced. Rather than speak for themselves, they usually sought leaders of the Party or prominent scientists to express their views. Thus, in 1970, Andrei Sakharov and two other scholars warned that the Soviet Union was in danger of reverting to "the status of a second-rate provincial power."[21] Military engineers sometimes sought the sponsorship of high officials in the Party to draw attention to their needs. There is also evidence that in the early 1980s some Soviet nuclear engineers agitated for more attention to safety in their industry. They believed that better containment structures around Soviet reactors would make the reactors more secure and also allow the Soviet nuclear industry to compete more successfully on the world market.[22] But despite these signs of a slowly awakening consciousness within the profession, the engineers were still remarkably passive compared to other professionals. No wonder engineering was sometimes called "the gray profession."

Peter Palchinsky at the exhibit he arranged for a manufacturing and mining trade fair in Turin, Italy, in 1911. Above the mine shaft entrance is the Russian mining engineers' symbol.

Nina Aleksandrovna Palchinsky at the time of her marriage to Peter, in 1899.

Peter Palchinsky, around 1909, when he was living in Western Europe and
working on his four-volume study of seaports. Photographs above his desk are
of ports and ships.

Peter and Nina Palchinsky, around 1916, after their return to Russia from exile in Western Europe.

The beginning of construction on the world's largest hydroelectric dam, Dneprostroi, on the Dnieper River.

Workers assembling a "snail," the water intake for a turbine, at Dneprostroi, 1932.

Col. Hugh L. Cooper, formerly of the U.S. Army Corps of Engineers, in
1930, standing before the great dam at Dneprostroi that he helped build.

Workers in 1929 digging at Magnitka Hill, site of the construction of the world's largest steel plant, at Magnitogorsk.

Young worker at Magnitogorsk, 1931.

Bricklayer at Magnitogorsk, 1931. The posters behind him say, "At full speed we will complete the Five-Year Plan in four years" and "We report that we are continuing the construction of the foundation of a socialist economy! The building of socialism in our country will be carried through to the finish!"

Building the blast furnace at Magnitogorsk.

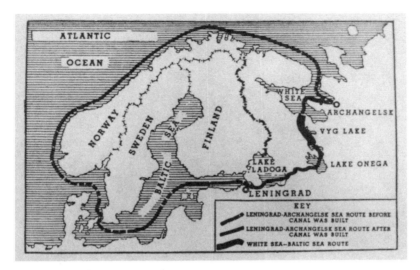

The White Sea Canal shortened the shipping route from Leningrad to the White Sea.

Prison workers at the White Sea Canal.

N. I. Khrustalev, imprisoned engineer and former "wrecker,"
who served as chief engineer for the building of the
White Sea Canal, 1932.

The first steamers on the White Sea Canal, 1933.

Construction of the Baikal-Amur Railway near Severobaikalsk, 1979.

Aerial view of the Chernobyl nuclear power plant after repairs had begun,
September 1986.

Monument erected where the White Sea Canal enters the White Sea.

5
Contemporary Engineering Failures

The industrialization projects of the First Five-Year Plan (Magnitostroi, Dneprostroi, Belomorstroi) violated the principles of sound and humane engineering practice advocated by Peter Palchinsky and his colleagues who had been trained before the Revolution. He participated in the discussions of the building of the great steel mill and hydroelectric dam, and his associates were the involuntary supervisors of the construction of the White Sea Canal. Their critique of these projects was based on direct experience.

In later decades most of the old engineers were off the scene, especially those who might say anything critical of orders that came from above, but their objections remained relevant. Soviet industrialization continued to be directed by a leadership that would not permit open debate of available options. Until the very end of the Soviet Union its industrialization program continued to follow a socially blind policy of fulfilling plan goals without examining the costs and alternatives. Three recent engineering blunders can serve as examples of this policy: the construction of a great railway in Siberia in the 1970s and 1980s, the Chernobyl nuclear accident of 1986, and the plight of the mining industry in the Don Basin at the end of the Soviet era.

The Baikal-Amur Railway

The largest construction project of the Brezhnev era was the building of the Great Baikal-Amur Railway (BAM), a two-thousand-mile freight and passenger rail line extending across southeast Siberia from the city of Novokuznetsk to the Pacific Ocean. Termed the "Project of the Century," its construction involved hundreds of thousands of workers and was constantly celebrated in Soviet newspapers, films, radio, art, television, and even novels and poetry.

The projected rail line went through some of the most difficult terrain on earth, over and through mountain ranges, swamps, and rivers. Its construction required working in the winter in temperatures so low that equipment failed and tools shattered. The project included the construction of 150 bridges, several over broad rivers such as the Lena and Upper Angara; 17 miles of railway tunnels; more than 200 stations and sidings; and over 20 towns and cities along the way. As one Soviet book published in English announced to its foreign audience, "The Baikal-Amur Railway outranks every other project in the history of railroad building anywhere in the world."[1]

The original promoters of BAM promised the unlocking of Siberia's hidden mineral riches and the flourishing of cities all along its route. Particularly important was access to the rich copper deposits of Udokan, in Chita province. The shipping of Siberian oil, along with coal and timber, to the Pacific Ocean for export was also a key goal. The plan's supporters spoke of creating a "mighty industrial belt along the BAM," including a major metallurgical complex.[2] Less often mentioned, but also a significant motivation for the project, was the wish to construct a railway that was protected from potential seizure by the Chinese. The older Trans-Siberian Railway runs near the Chinese border for hundreds of miles, while the BAM would be farther north in a more secure area.

Leonid Brezhnev announced the beginning of BAM's construction in 1974. He described it as a continuation of the tradition of "labor accomplishments by our people" such as the Dnieper hydro-electric station and the steel city of Magnitogorsk.[3] BAM was indeed in that tradition, but it also demonstrated that the Soviet Union had changed since the days of the early Five-Year Plans. Under Brezhnev, prison labor was not used for the construction of BAM (although years earlier prisoners had, under Stalin, begun portions of the line). Far more attention was paid to the workers' living conditions than had been done in the earlier projects, though life at the remote construction sites remained spartan.

Most laborers on BAM were volunteers, attracted to Siberia by offers of high wages and even, on occasion, the promise of a small automobile (the Zhiguli, a variant of the Fiat) after three years of work in Siberia. The project became a campaign of the Young Communist League, the Komsomol, which organized brigades of young men and women from all over the Soviet Union to spend a few years working on BAM. The poet Iurii Razumovskii called on Soviet youth to join the campaign:[4]

There's only mountains here today
And marshland and taiga,
A wilderness run over by
The bear and *kabarga*.
But soon by humpback bridges
The river will be crossed
And stations in the forest
Will take their timeless post . . .
We'll have a jolly time yet, friend,
And sing and dance and all,
When the Pacific Ocean
We've linked with Lake Baikal.
Along this very railway line
Over rivers broad and free
We'll speed along by special train

To the next century.
You're neither invalid nor old,
Then why, lads, wait so long?
To the project of the century
The country calls its young.

A few of the youths embarked for Siberia in a spirit of dedication and enthusiasm, more because they were interested in the relatively high wages, some because they had been unsuccessful in finding career advancement at home. Many of them were organized in ethnic units, enabling the Communist Party activists who oversaw the work to organize "socialist competitions" among the various nationalities for fulfillment of work quotas. Thus the Armenians would compete with the Ukrainians, the Russians with the Uzbeks, the Georgians with the Belorussians, and so forth.

The construction of BAM was the last gasp of the old Soviet methods of organizing work by hortatory campaigns, with minimal attention to the difficult technical and social issues considered so important by Palchinsky and his colleagues. And BAM certainly did not accord with Palchinsky's motto of "achieving the maximum social benefit with the least effort." The railroad project was organized like a military campaign that had to be successfully completed in a hurry, at any cost. But the tolerance of the Soviet people for such ideological programs had greatly declined since the thirties. The workers had been urged on a few times too many.

The youths who went to Siberia in the seventies and eighties were better educated than their predecessors who worked on the grandiose projects of the early Five-Year Plans, and they were more sophisticated and skeptical. Some had become cynical about Soviet goals. Most of them knew at least something about the bloody purges described to a Party meeting by Khrushchev in 1956 in a secret speech that soon leaked out, most had heard of the invasions by Soviet troops of Hungary in the same year and of Czechoslovakia in 1968, and a few had participated in dissident activities during the sixties and seventies, or had friends or family members who had done so. All of them knew that after years of promises of plenty the Soviet Union was still a land of "deficit" commodities, a place where ordinary living was a difficult task. While most remained loyal in some way to the Soviet Union, they were beginning to ask questions. They returned home from the BAM construction sites with stories of waste and excess, and of mindless exhortation. Enthusiastic poetry no longer could harness a new generation of Soviet youth.

While the construction of BAM was not the scene of horrors that the building of the White Sea Canal had been, it nonetheless displayed many of the same profligate characteristics of all large Soviet construction projects since the early thirties. The decision to build the railway was not an open process in which defenders and proponents of the project or its different variations could publicly present their views, with supporting evidence; instead the decision was made by the leaders of the Communist Party, relying on a small circle of technocratic advisors nurtured on the familiar ethos of extensive development of production facilities without regard for costs—economic, environmental, or social. Indeed, the construction of BAM was already in full swing by the time the design of the project was finally approved, in 1977.[5] The locations of many sections of the line were determined as the building proceeded, not in advance. One can well imagine what Palchinsky and his like-minded engineering associates would have said about such an ill-founded and hasty approach.

Once the decision to build the line had been made, anyone who voiced criticism was immediately branded a skeptic, a laggard, a person lacking enthusiasm for the construction of communism. Such a person was not sent to prison, as might have happened in the thirties, but was simply denied publication by the government-controlled media. The Soviet propaganda organs exalted BAM as a breathtaking challenge requiring wartime fervor, perseverance, and engagement. The enemy was not a hostile army but nature itself—the frozen taiga in the winter, and permafrost and mosquito-ridden swamps in the summer. In the face of these obstacles, success was achieved through sheer determination. There was no place for carpers and critics.

Despite the exhortations of the government and the Communist Party, the BAM project soon fell far behind schedule and grossly exceeded the inadequate cost estimates. The supervisors naturally looked for ways to speed construction and cut costs. One way was to use lighter rails than planned. In the late seventies and early eighties light R-50 rails were used on several sections of the track, instead of the hardened and heavier R-65 rails.[6] The result: three train wrecks before the railway was even completed, and the premature deterioration of the rails, especially on curves. Eventually all the light rails were replaced, but only after the supervising engineers falsely informed their superiors that the track had been completed as specified. Safety and the long-term budget were sacrificed to the time schedule and the short-term budget, a typical result of the accelerated tempos of Soviet projects.

Another way to hasten construction and save money was by using military troops in construction brigades. The employment of soldiers in construction was a traditional practice in the Soviet Union. Soldiers helped build the Ignalina and Gorky atomic power stations and many of the irrigation canals along the Volga River. Indeed, they could be seen on almost every Moscow street, repairing roads and putting up buildings. Whenever a minister or Party functionary experienced difficulty in getting a certain project finished on time, he would appeal to the Ministry of Defense, saying, "Unless you give us soldiers, the plan is lost."[7] The builders of BAM were no different. They called for help from the military authorities, who complied. Much of the eastern section of the railway line was built by soldiers, who also did 25 percent of the heavy work, such as excavation and blasting, along the entire line.[8]

The employment of soldiers for these tasks was ethically and economically flawed. Although they were not prison laborers, as their predecessors often had been, they were definitely involuntary workers, ordered to their work sites by their commanding officers. They were assigned the most unpleasant jobs, the ones the voluntary workers did not wish to perform. Furthermore, they were paid very little, since they were drafted soldiers of the lowest ranks. Their monthly salaries were a fraction of those of the voluntary workers doing less onerous tasks. Military laborers were logical descendants of the prison laborers of the past. Under Brezhnev the Soviet Union was a less cruel place than it had been under Stalin, and soldier-laborers were treated better than prisoners. But they performed a similar function: they would silently comply with orders and they could be thrown into the breach whenever a construction schedule was falling behind.

During most of the construction of BAM the use of soldiers as laborers was not discussed. After the advent of Gorbachev's *perestroika*, however (and before the completion of BAM), the Soviet press began to reveal the extent to which soldiers were being used. One writer in *Pravda* castigated the practice on ethical grounds and also rightly noted that it distorted economic evaluations of construction: "The use of soldiers to do construction work in the civilian economy has long been a kind of 'sacred cow' that could not be criticized in the press. But the fact is that using soldiers to fulfill plan assignments 'corrupts' many of our departments, for their desires are no longer constrained by their resources—after all, their manpower costs them nothing."[9]

The employment of soldiers and short-sighted cost-cutting measures did not save the plan. The completion of BAM had been originally scheduled for 1983, but that goal appeared increasingly unrealistic. A particularly difficult section of the line was the Northern Muia Tunnel in Buriatia, at 9.5 miles the longest underground section on the route. Before the railway line had even been designed, geologists and engineers in Buriatia had warned the Soviet government that the intense seismic activity in the area made a tunnel inadvisable, and recommended instead a long bypass of the area.[10] Unwilling to accept the time delay the bypass's construction would cause, the government overruled the specialists and approved the tunnel.

Building the tunnel turned out to be much more challenging than expected. Pressured by their bosses to make 1984 the year to drive the "golden spike" symbolizing BAM's completion, the supervising engineers desperately threw together a 17.4-mile bypass that included grades too steep for regular freight trains to negotiate. A newspaper correspondent who made the trip observed that "only a slalom skier could handle this road."[11] Nonetheless, the Soviet press could announce in 1984 that BAM had been completed. In truth, five more years would pass before the first regular freight trains could use the length of BAM, and even then problems remained. The second temporary bypass around the fragile tunnel also allowed only slow and limited traffic.

The construction of the great railway under such rushed conditions did major damage to the environment of Siberia.[12] The sewage produced by the hordes of construction workers had no place to go since in the winter the creeks and small rivers often froze to the bottom and year-round the soil remained frozen a foot or two below the surface. Rivers and streams along the construction line were heavily polluted with oil, grease, rubbish, and discarded equipment. In the winter the diesel engines of the heavy equipment were allowed to run all night, since otherwise they could not be started in the cold of the morning, creating pervasive smog and pollution. Damage to the tundra of the region, which was exceedingly fragile, exposed the thin soil, which would not recover for decades, if ever. The indigenous Siberians despised the Russians for killing the local game, ruining the land and water, and then leaving for home as soon as they had spent enough time on the site to earn their Zhiguli automobiles.[13] Lake Baikal, the oldest and deepest lake in the world, home to many unique species, became a major supply route for BAM; the lake abounded with transport boats in summer and trucks on the ice in winter. The newly awakened environmental movement in the Soviet Union vehemently protested the abuse of the lake.

BAM was operational by the early nineties, but its economic value remained dubious. The original hope that oil for export would be the most valuable freight on the railway was not realized, because drilling for oil in that area of Siberia was so difficult that it was indefinitely postponed. The development of the copper industry in Udokan, the second most important economic motivation for the railway, had in the meantime also been postponed. With the slow-down of the Soviet economy, and the eventual dissolution of the Soviet Union itself, no large-scale mineral extraction from the area has been scheduled. So far the only significant cargo shipped the length of BAM has been timber, although it is more expensive than timber from elsewhere in Russia. Defenders of BAM often cite the profitable trade in coal from Southern Yakutia; that coal does not travel on BAM itself but on a spur line connecting to the old Trans-Siberian Railway. A Russian economist observed in 1988: "For now there is nothing to haul on the new and expensive railroad, and the BAM is an unprofitable venture. We must take care that underutilization of the line over the next few years does not encourage neglect of necessary maintenance, postponement of infrastructure construction and other false economies that could have negative consequences for decades."[14]

The Significance of Chernobyl

The beginning of the former Soviet Union's recovery from narrow technocratic visions probably dates from the catastrophe at the Chernobyl nuclear power plant in 1986. The Chernobyl explosion has been analyzed extensively in Western, Soviet, and post-Soviet publications. With few exceptions, however, the authors have concentrated on the technical details of the reactor itself, the behavior of its supervising personnel, and the resulting social and economic damage. Rarely have analysts noted that Chernobyl was a product of the standard Soviet industrialization policy.that emphasized gigantic projects over smaller ones, centralized plans over locally sensitive ones, output above safety, technology above human beings, closed decision making to the detriment of critical debate, and, above all, a madly rushed tempo.[15] If nuclear power was different from the rest of Soviet industry, it was only because the possible consequences of failure resulting from standard policy were especially catastrophic. Designed, constructed, and operated within the framework of the traditional Soviet approach to industrialization, Chernobyl was a disaster waiting to happen.

Soviet leaders constantly urged all forms of industry to expand as rapidly as possible so that the USSR could become the most powerful country on earth. Nuclear power was no exception; the Tenth, Eleventh, and Twelfth Five-Year Plans (1976–1990) called for dramatic growth. By the end of 1980 there were to be a total of twenty-four reactors in operation, of which thirteen were of the graphite-moderated type built at Chernobyl (RBMKs), a design rejected in almost all other countries because of its inherent instability. Ten others were the somewhat safer light-water pressurized reactors (VVERs). The remaining one was a fast-breeder reactor at the Bilibino station. After 1980 new reactors were to be steadily brought on-line; between 1986 and 1990 the generation of nuclear power in the USSR was to increase by 250 percent, for a total of 69,000 megawatts.[16] In 1986, when the accident occurred, the USSR had 43 operating reactors, 36 others under construction, and another 34 reactors in the planning stage.[17] Although the Soviet Union decided even before the Chernobyl event to shift gradually from the obsolescent RBMK to more modern designs, even as late as the winter of 1992–93, almost seven years after the Chernobyl accident, fifteen

Chernobyl-type RBMK reactors were still in operation on the territory of the former Soviet Union. Senior nuclear officials announced that, because of dire need for electricity, the RBMK reactors would go on operating indefinitely.[18]

Soviet policy for nuclear reactors was to build very large ones (usually around 1000 megawatts), and to site as many as six of these reactors in one location, thereby creating enormous power centers. Four 1000-megawatt reactors were already in operation at Chernobyl in 1986 and two more were under construction. Larger reactors, including a 2400-megawatt RBMK model, were under discussion.[19]

To produce reactors rapidly, the Soviet Union established an assembly-line plant for their manufacture, the Atommash plant at Volgadonsk. An American analyst of Soviet nuclear power, Paul Josephson, observed, "Planning and standardization have reached the stage where the Ministry of Power and Electrification published manuals for reactor construction detailing how to prepare sites, what bulldozers, earthmovers, and trucks to use, cooling tower construction, and what cranes to employ when building largely standardized reactors."[20] This uniform policy neglected the unique characteristics of each site, such as groundwater levels, population density, type of soil or rock foundation, and seismic activity, factors that affect both the decision to build a reactor on a certain site and the choice of construction methods.

Before Chernobyl almost no criticism of nuclear power was permitted in the Soviet popular press. The most frequently cited exception was a 1979 article coauthored by a physicist and an economist in the prominent journal *Kommunist.* They questioned the current siting policies for Soviet nuclear power plants and urged that future power plants be placed in remote areas, far from large cities.[21] Yet even they supported the expansion of nuclear power. And their fears about safety were soon repudiated in the Soviet press. Members of the scientific, engineering, and managerial elite published articles reiterating the importance of nuclear power to economic expansion and dismissing safety concerns. In 1980 Academician M. A. Styrik-ovich wrote in a popular Soviet magazine, "Nuclear power stations are like stars that shine all day long! We shall sow them all over the land. They are perfectly safe!"[22]

Many of the specific flaws and mistakes that led to the explo-sion at the Chernobyl Reactor Unit No. 4 on 26 April, 1986, can be linked to the general characteristics of Soviet industrial policy. The failures at Chernobyl include the lack of adequate containment of the reactor, the continued use of an obsolescent reactor that was particularly difficult to control, a low level of qualification and train-ing of the supervisory personnel, and a general disregard for safety. The pressure to increase output of electricity also contributed to the disaster, which occurred during an experiment to see if more elec-tricity could be squeezed out of a reactor as it was being shut down for maintenance.

Rather than an accident, the Chernobyl explosion might be more appropriately regarded as the consequence of a series of policy decisions. Nuclear power is inherently a dangerous technology, in whatever country it is used, and the former Soviet Union is not unique in having had nuclear accidents or in taking risks. Nonethe-less, Soviet policy on nuclear power converted a risky industry into a terrifying one. The especially dangerous characteristics of nuclear power plants in the former Soviet Union became even clearer in March 1992, when another RBMK reactor of the Chernobyl type leaked radioactive gas near St. Petersburg. Many West Europeans demanded that all reactors of this type be closed down immediately. Klaus Toepfer, the minister of the environment in Germany, stated, "We remain convinced that the RBMK reactors cannot be brought up to standards and that they must be shut down as fast as possible."[23]

The calamity at Chernobyl was soon followed by a series of less spectacular but also significant technical failures involving nu-clear submarines, transportation (ship and train wrecks), and envi-ronmental disasters. Mikhail Gorbachev, trained as a lawyer, not as an engineer, identified their cause as "the human factor," using the same term Palchinsky did years ago. Gorbachev began calling for new approaches to technology, with much more attention to contextual and social issues such as economics, safety, workers' benefits, environmental risks, and managerial practices that took into account psychological and sociological factors. He turned more and more to economists, sociologists, and even historians for advice—and less and less to engineers.[24] The ghost of Peter Palchinsky, who had warned of the effects of narrow technical education and ignorance of social issues, had returned to haunt the Soviet Union.

The Don Basin Revisited

In July 1989 the first widespread strike in Soviet history broke out among Siberian coal miners and spread to the coal miners of the Don Basin, where over 250,000 miners left their jobs. The strike was a watershed in Soviet labor history. Under the new conditions of *glasnost'* the grievances of the miners were made public, and social scientists were allowed to study the working and living conditions of the Don Basin miners. Peter Palchinsky hovers over these studies, for the centers of the 1989 strike movement were in the same area and sometimes at the same mines that Palchinsky had studied almost ninety years earlier. Of 121 mines in the Don Basin in the late 1980s, 36 had been in use for over 70 years and several for over 100 years.[25] It was the conditions of the Don Basin mines at the turn of the century that had radicalized Palchinsky and launched him, as a young engineer, on a path of social and economic reform. Sixty years after his death the new studies of Don Basin miners showed that although the workers were better educated and were provided with somewhat better housing than their tsarist predecessors, they still were deprived of basic necessities and continued to live and work in appalling and miserable conditions.

In the post–World War II period the richest and most accessible veins of coal in the Don Basin were exhausted, forcing the mine supervisors to order excavations deeper and deeper into the earth. By 1985 one mine had reached a depth of 4,326 feet. Under the particular geological conditions of the Don Basin, at these depths the danger of explosion from methane gas accumulation is intense. Already in 1976, in one of the first scientific exchanges between the USSR and the United States on science and technology policy, the Soviet members proposed joint research on methane gas explosions, a concern in American mines as well as Soviet ones.[26] The two sides were able to exchange valuable information on standard ways of preventing such explosions: forecasting the onset of gas bursts, degassing outburst-prone coal beds, installing gas sensor and alarm systems, providing adequate ventilation and temperature control in the mine shafts, establishing loudspeaker warning systems and good communications throughout the mines, and drawing up emergency evacuation procedures. In 1976, however, the Soviet participants in the exchange would give no statistics on the average number of casualties annually from such explosions. In the United States the annual number of deaths from all causes in coal mines in the mid-1970s was about 140, of which less than 10 were from gas explosions.[27] After the strike of 1989, it became clear that in many Soviet mines ventilating equipment was either absent or unreliable, and that communication systems were often similarly lacking.

We still do not have reliable statistics on deaths and injuries in the Don Basin mines, which produce about a third of the former Soviet Union's coal, but in the wake of the 1989 strike we learned that 44 miners died (from all causes, not just gas explosions) in July 1989, 67 in August, and a total of 431 in the first eight months of 1989. A. S. Dubovik, a strike leader in the Don area, maintained that three or four lives were lost for every million tons of coal extracted. (The comparable figure in the United States is about .2 lives per million tons.)[28]

Theoretically, Soviet miners were given many advantages over other laborers and even enjoyed privileges not found in Western countries. According to the Soviet government, in the 1980s miners in the Don Basin were paid an average of 350 to 400 rubles a month, almost double the average wage in Soviet industry.[29] The Don Basin miner supposedly worked a five-day, thirty-hour week, was eligible for retirement at age fifty, enjoyed free medical care from birth to death, and received a retirement pension of up to 170 rubles a month.[30] In actual fact, however, the workers in the Don area in the 1980s were under heavy pressure from their supervisors to increase output and as a result worked far more, on average, than thirty hours a week. Travel time from the shaft to the coal face could be as long as an hour each way, but was not counted as work time. Saturdays, supposedly a day off, became a normal workday, and increasingly Sundays as well, with the additional time paid at the standard rate, rather than overtime.[31]

The miners and their families had great difficulty finding adequate food and supplies. Shortages of meat, fruit, and vegetables were endemic. Most irritating of all to the miners was the frequent absence of soap and soap powder, which made washing up after leaving the mines each day almost impossible. Here was a humiliating deprivation not mentioned even in the complaints of miners in tsarist times to Palchinsky. Indeed, the demand for adequate supplies of soap occupied a major place in the grievances of the Don miners during the 1989 strike.

After the strike, sociologists from the Ukrainian Academy of Sciences asked 216 people (199 workers and 17 engineers) the main reasons for the strike. The results of this poll were not compiled very rigorously, but they give a rough idea of the areas of discontent (in percent; respondents gave multiple answers): Shortages of basic supplies, 86 percent; low wages, 79; brevity of vacations, 62; inadequacy of pensions, 56; high prices of supplies, unsatisfactory housing conditions, and poor relations with the administration, 39–41;

poor working conditions, 33; lack of social justice, 32; poor medical services, 25.[32]

Most shocking of all, at least to outside observers, were the conditions of the miners' housing. Although that housing had improved since Palchinsky inspected Don miners' barracks at the turn of the century, the difference three generations later was dismayingly small. One miner observed after the 1989 strike, "Many miners still live in such conditions that if a film were shot here, you would think it was 1905."[33] At the beginning of 1989, according to an official report issued after the strike, 63 percent of the miners' residences lacked hot water, 20 percent did not have even cold water, and 26 percent were not connected to a sewage system. Approximately 17 percent of the miners had no homes or apartments at all, and lived with friends or in barracks.[34]

Journalists who interviewed the Don miners after the 1989 strike were told that the suffering of the miners was rooted far back in Soviet history. The miners reported that immediately after the Revolution, in the early twenties, miners, like other Soviet workers, had been promised a new era in which they would live better than workers anywhere else in the world. The end of the time when it seemed possible that such promises would be fulfilled, they said, came with the terror and repression in the late twenties. In 1928 the head of the trade union organization, M. P. Tomsky, was forced out of his office. (Not coincidentally, Palchinsky, the most outspoken of the engineers sympathetic to social issues and workers' conditions, was arrested in the same year.) From that time forward, the miners reported, the "interests of the workers were neglected."[35] The coal miners attempted a strike under Khrushchev, when political conditions momentarily relaxed a bit, but their uprising was brutally crushed by armed troops. Not until the era of Gorbachev's reforms did they have the courage to strike for improvements in their living and working conditions.

Soon after the strike began, Gorbachev appeared on television and acknowledged most of the miners' grievances. Subsequently, the government met with the leaders of the strike committees and agreed to fulfill their twenty-five demands. The array of promised reforms included wage and pension raises, a guarantee of Sundays as rest days, a minimum of 800 grams of soap per worker per month, increases in food and supplies, improvements in safety, and an assurance of apartment housing for all miners. In time, many of these promises were broken; the increased wages and pensions quickly became inadequate in the face of inflation. Worker unrest continued. Nonetheless, the miners' strike of 1989 was a signal event, the first successful large strike in Soviet history.

In an essay in 1926 Peter Palchinsky wrote that the active workers of Soviet Russia, eighty million strong, were a "non-utilized force compared to which all the other great natural riches of the country paled in significance."[36] He believed that Russia could become not only a great power but a humane civilization if only it would combine wise use of its rich natural resources with care for that workforce. Long before the miners' strike of 1989 the Soviet Union had become a military superpower and possessed one of the mightiest heavy industrial establishments in the world. The victory of industrialization, however, turned out to be entirely hollow, for in achieving it the leaders of the state had lost the support of their own population.

Epilogue: The Ghost of Peter Palchinsky

In 1992, the playwright and president of independent Czechoslovakia Vaclav Havel wrote that the fall of communism marked the end of an era, the demise of thought based on scientific objectivity. Marxism, he believed, was committed to "arrogant, absolutist reason," and its failure meant that instead of relying on objectivity a person must "trust in his own subjectivity as his principal link with the subjectivity of the world."[1]

What would Peter Palchinsky say in response to these words? He would defend as stoutly as Havel their shared goal of a humane world, but he would laugh, perhaps through tears, at the description of Soviet-style Marxism as the pinnacle of scientific rationalism and objectivity. Was the building of the White Sea Canal in the wrong place and by the most primitive methods, at the cost of hundreds of thousands of prisoners' lives, the blossoming of rationality? Was the disregard of the best technical specialists' advice in the construction of Magnitogorsk, the Dnieper dam, and the Baikal-Amur Railway a similar victory for objectivity? Was the education of the largest army of engineers the world has ever seen—people who would come to rule the entire Soviet bureaucracy—in such a way that they knew almost nothing of modern economics and politics an achievement of science? What was Stalin's imperious demand for industrial expansion at a rate that was technically unfeasible and shockingly wasteful of human lives, if not a flight of rank subjectivity? And even long after the death of Stalin, into the 1980s, what was the Soviet insistence on maintaining inefficient state farms and giant state factories, if not an expression of willful dogmatism that flew in the face of a mountain of empirical data worldwide about economic structures that were more efficient and more just?

Palchinsky could confirm that these irrational characteristics in Soviet communism did not first become visible in Havel's time, but had been discerned and described in the 1920s by himself and many others. Their critiques were put forward in the name of scientific rationality and social justice, principles that Palchinsky hoped to combine. His criticisms of early Soviet industrialization policies help us to understand the later failure of the Soviet Union to sustain its drive toward modernization.

The initial industrialization projects of Soviet history were buoyed by the social energy brought by the new Communist government to workers accustomed to the ravages and oppression of the tsarist regime. The Communists issued to the workers a promissory note that the new Soviet society, no matter how difficult at first, would eventually be bountiful and humane. The note was good for one generation. Many of the workers who labored under the dreadful conditions of Dneprostroi, Magnitostroi, and the other industrialization projects of the first Five-Year Plans managed to retain their faith that the future would bring a materially and spiritually rich life. Although doubts began to grow by the end of the thirties amid the oppression and violence of the purges and the broken promises to the workers, World War II provided the government with a reprieve. Nationalistic sentiments energized a great industrial and military effort that was successful and in which the state and its citizens took justifiable pride. But the destruction inflicted on the Soviet Union by Nazi Germany was severe, and recovery took many years. The Soviet government and the Communist Party justified the continued deprivation in the immediate postwar years by emphasizing how much the Soviet Union had suffered. Quite a few Soviet citizens found the argument persuasive, for they knew from personal experience how difficult the war had been.

By the late sixties and early seventies the ideological promise of a coming socialist utopia or the legacy of past wartime destruction could no longer be invoked to justify present suffering. When Brezhnev called on young workers to help build the gigantic Baikal-Amur Railway, he could no longer rely on fervor for a new society. Hortatory sentiments about building communism had lost their appeal to the Soviet worker. Instead, the Soviet government sought to motivate workers to go to Siberia through the promises of higher pay and Zhiguli automobiles. The material incentives that the Soviet system could offer in the seventies proved far less compelling than the dream of a socialist society, now forever lost.

The erosion of faith accelerated as the citizens of the Soviet Union became increasingly aware that although their country had become a great industrial power, their standard of living matched that of third-world countries. By the seventies the Soviet Union was the largest producer in the world of steel, lead, asbestos, oil, cement, and several other basic industrial goods. But the cost in human and environmental terms of a blind fixation on output was perilously high. Food and consumer goods were often unavailable because the political bosses insisted on producing steel for heavy industry and the armed forces. Life expectancy declined until the Soviet Union ranked thirty-second in the world. Infant mortality rose until the Soviet Union ranked fiftieth in the world, after Maritius and Barbados.[2] The environment was a disaster, especially around industrial cities like Magnitogorsk and in areas requiring irrigation, such as Central Asia.

The workers responded to this neglect of their needs by lapsing into apathy. Their naive hope that, in time, the Soviet regime would keep its promises took a long time to fade, but eventually it disappeared completely. In the last years of the Soviet Union the attitude of the members of the proletariat, the supposed beneficiaries of communism, was expressed in the cynical observation "We pretend to work and they pretend to pay us." In the final years of the regime, the phrase was revised to "They pretend to rule and we pretend to obey." Against this background, Palchinsky's advice that the "human factor" should be utmost in the mind of the engineer or manager was prescient. The gross neglect of human beings by the Soviet regime was a primary reason that it collapsed so strikingly easily. In the end, it had almost no defenders.

An aspect of Palchinsky's vision of the relation of technology and society that deserves reiteration was its ethical sensitivity, especially compared to technocratic doctrines being promoted elsewhere during his time. While American engineers and their followers in other countries praised "Taylorism" and "Fordism" for their ability to increase productivity, Palchinsky asked what effects the imposition of these methods might have on workers. As a person for whom the workers' welfare had always been uppermost, he was not willing to accept efficiency or productivity as the only legitimate goals of industry. In his vision of society, justice and efficiency could be made to work in tandem, not in opposition. This view is strikingly similar to that of industrial managers of recent times who have tried to improve the productivity of assembly lines and shop floors by humanizing the work environment. Yes, Palchinsky was a technocrat, and he possessed many of the shortcomings of technocrats, including some of those eloquently criticized by Havel, but he was one of the most socially aware technocrats the world has seen. Would that most of the graduates of engineering institutions, and not only in Russia, possessed his breadth of understanding.

Palchinsky's vision of a rational and just relation between technology and society was immensely superior to the Soviet Marxist dogma that succeeded him, and even to Western engineering doctrines of the time. Yet Palchinsky's conviction of the superiority of government ownership of land and factories seems unpersuasive today. We have learned that the concentration of economic power in the state's hands is dangerous and inefficient. Palchinsky hoped that in a socialist society, in the absence of capitalists and entrepreneurs, engineers would run the economy, an aspiration that failed to appreciate the diversity of talents and sensibilities needed by political leaders, managers, workers, and administrators. In his preference for engineers Palchinsky displayed a flaw identified by Havel, even if Havel exaggeratedly pinned the blame on scientific thought itself.

Although given special impetus by the Russian Revolution, this granting of preeminence to engineers was not unique to Russia in the twenties. It was a part of a widespread movement that one American historian has called "the revolt of the engineers."[3] In the United States, Germany, France, and Great Britain, similar ideas were expressed by professional engineers. The movement may have achieved its peak in 1928, when a mining engineer, Herbert Hoover, was elected president of the United States. Within a few years the movement had run its course and engineers settled back into their traditional roles as paid employees or consultants to large companies or government agencies, or, much more rarely, entrepreneurs who became capitalists themselves.

Palchinsky accurately predicted the harm caused by hasty industrialization projects that violated both good engineering practice and normal ethical standards. His questions about Soviet industrialization continue to haunt the former Soviet Union and cast doubts on what many regard as their greatest achievements. His belief that efficiency and justice must always be linked is one that haunts industrial civilization everywhere, from the steel city of Magnitogorsk, the failed garden of socialism, to its erstwhile model, Gary, Indiana, plagued with the familiar American problems of poverty, unemployment, drugs, and urban blight.

Because of his prescience, some may wish to call Palchinsky a prophet. Less convinced of the existence of revealed truth, I prefer to call him a ghost. That ghost haunts most pervasively the inefficient, polluted, and inhumane industrial cities of the former Soviet Union, since Palchinsky's original criticisms were directed toward industry in his native land, but it hovers uncomfortably over the industrial wastelands of other countries as well.

Where did Palchinsky and his associates obtain their vision of the humane engineer, the liberally educated specialist as much at home with economic and social issues as with a calculator or slide rule? Not from their engineering colleagues in the West, nor from Marxism, but from their own experiences. Most of them had opposed the tsarist regime and some of them, like Palchinsky, had been associated with radical politics. They criticized the tsarist government for its inattention to workers' needs, for its unwillingness to create a pluralistic society, for its fear that economic development would bring pressures for a democratic form of government. Some of them, again exemplified by Palchinsky, had been imprisoned for their political views long before the Russian Revolution of 1917. As a result of their experiences, before the advent of the Soviet regime they already considered engineering problems as intrinsically linked to social and political issues. That link had been obvious in their own lives; Palchinsky had seen it clearly in his studies of the Don Basin miners at the beginning of the century.

It was probably not an accident that Palchinsky was a mining engineer rather than another type of technical specialist. Mining engineers are accustomed to working at very remote sites, often previously uninhabited, and they are frequently responsible not only for planning the mines but also for their day-to-day operation. They know that the problems facing them are as much social as they are technical. Without the construction of communities, including housing, schools, hospitals, transportation networks, and recreational facilities, the ore cannot be obtained.[4]

After the Bolshevik Revolution, Palchinsky and his associates remained loyal to the concept of the socially engaged engineer. They saw novel possibilities for engineers to be social planners as well as technical consultants; under socialism the industrial communities that they would plan would be far superior to any that had grown up around factories and mines under capitalism. This ambition to be at the heart of social planning collided head-on with Stalin's determination to concentrate power in his own hands. He accused the engineers of treason, when all that they were guilty of was trying to increase their influence. The assault of Stalin's secret police on the leading engineers and their professional associations was so violent that they remained silent politically until the end of the Soviet Union.

In the decades beginning with the thirties the Soviet Union produced more engineers than any other country in the world, but these new engineers were ones who had learned the lesson that they must not meddle in politics or social issues. Even if they had wished to do so, their education was so narrow that they had little preparation for such problems. Under Khrushchev and Brezhnev they gradually rose to positions of influence in the Soviet government and in the Communist Party, but Palchinsky would have been horrified by the type of engineer who came to occupy most of the top positions in Soviet society.

One of the great ironies of the story of Peter Palchinsky is that in the end he failed to follow his own precepts. He always urged engineers to look at engineering problems in their social and political context. By the end of the twenties the political context of Soviet Russia was changing dramatically in a way that made Palchinsky's policy recommendations unrealistic for the country and dangerous for him. Yet Palchinsky persevered as if the political situation had not changed. His call for an independent and influential profession of engineering had no chance of success in the Soviet Union controlled by Stalin, who permitted no threat to his authority from any direction. Palchinsky's behavior became more and more reckless as he challenged Stalin's approaches to industrialization. It was a recklessness that one can admire, even if one sees that it was foolhardy.

Palchinsky's stubborn determination—inconsistent as it was
with his lifelong insistence that engineering problems be solved
within their social and political context—probably influenced even
his death. Alone among the engineers accused of conspiracy and
treason in the Industrial Party Trial, he was not brought before the
court but instead was executed in secrecy. All the others publicly
confessed to the false charges, solemnly affirming before the court
that they were the agents of capitalist powers and had conspired
to overthrow the Soviet Union. Because they confessed they were
imprisoned instead of being executed. It is quite probable that Palchinsky's execution resulted from his refusal, even under torture, to
confess to crimes he did not commit.[5] Palchinsky always prided
himself on being a rational engineer. One can question whether his
final act was rational, but one cannot question its bravery.

Notes

1. THE RADICAL ENGINEER

1. Throughout this book I spell his name "Peter Palchinsky" even though the Library of Congress system of transliteration used elsewhere in the book would dictate "Petr Pal'chinskii." The first form is the way in which he was known in most countries abroad and it is less distracting to the English-language reader.

2. "Ot Ob"edinennogo Gosudarstvennogo Politicheskogo Upravleniia," *Izvestiia* (May 24, 1929), p. 1.

3. Aleksandr Solzhenitsyn, *The Gulag Archipelago*, vol. 1 (New York: Harper and Row, 1974), p. 6.

4. Samuel A. Oppenheim, "Pal'chinskii, Petr Akimovich," *The Modern Encyclopedia of Russian and Soviet History*, ed. Joseph L. Wieczynski, vol. 26 (Gulf Breeze, Fla.: Academic International Press, 1982), pp. 188–189.

5. Robert Campbell, *Soviet Economic Power: Its Organization, Growth and Challenge* (Cambridge, Mass.: Houghton Mifflin, 1960), pp. 51, 54–55.

6. Palchinsky served as the inspiration for a fictional character, P. A. Obodovsky, in Aleksandr Solzhenitsyn's multi-volume novel of the Russian Revolution, *The Red Wheel*. See *August 1914: The Red Wheel I* (London and New York: Penguin Books, 1989), pp. vi, 755–767, and *Sobranie sochinenii* (Vermont and Paris: YMCA Press, 1984–1987), esp. vol. 13, pp. 376–480; vol. 15, 477–480; vol. 16, 216–219; and vol. 17, 430–433.

7. The Central State Archive of the October Revolution, hereafter TsGAOR, f. 3348, op. 1, d. 1010, l. 4.

8. TsGAOR, f. 3348, op. 1, d. 1010, l. 3; l. 8.

9. TsGAOR, f. 3348, op. 1, d. 1010, l. 45.

10. Ibid., l. 60.

11. Ibid., l. 28.

12. "*Curriculum vitae* gornago inzhenera Petra Ioakimovicha Pal'chinskago," TsGAOR, f. 3348, op. 1, ed. khr. 3, l. 1.

13. *Katalog knig kazanskoi biblioteki A. I. Pal'chinskoi* (Kazan, 1896).

14. Letter from Julia to Peter, 12 January 1909, TsGAOR, f. 3348.

15. TsGAOR, f. 3348, op. 1, ed. khr. 525, l. 1.

16. For graphic descriptions of the miners' working conditions, see Charters Wynn, *Workers, Strikers, and Pogroms: The Donbass-Dnepr Bend in Late Imperial Russia, 1870–1905* (Princeton: Princeton University Press, 1992), pp. 67–94, and Aleksandr I. Fenin, *Coal and Politics in Late Imperial Russia*, trans. Alexandre Fediaevsky, ed. Susan P. McCaffray (DeKalb: Northern Illinois University Press, 1990).

17. P. I. Palchinsky, "Zhilishcha dlia rabochikh na rudnikakh Donetskogo basseina," *Gornyi zhurnal*, 9 (September 1906).

18. See Paul Avrich, *The Russian Anarchists* (Princeton: Princeton University Press, 1967).

19. Palchinsky cited, in particular, Kropotkin's books *Bread and Will* and *Fields, Factories, and Workshops*; TsGAOR, f. 3348, op. 1, ed. khr. 595.

20. Palchinsky, "Nekotorye dannye po rabochemu voprosu na Kamennougol'nykh rudnikakh Cheremkhovskago kamennougol'nago raiona v Irkutskoi gubernii" (Irkutsk, 1903); "Zhilishcha dlia rabochikh na rudnikakh Donetskago basseina," *Gornyi zhurnal* 9 (1908); "Vos'michasovoi rabochii den' na rudnikakh Frantsii i znachenie ego dlia kamennougol'noi promyshlennosti i vsego ekonomicheskago polozheniia strany," *Gorno-Zavodskii listok* 53 (1908); "Reforma Avstriiskago sotsial'nago strakhovaniia i vopros o strakhovanii rabotnikov gornago dela," *Gornozavodskoe delo* 33: 1–2 (1913), pp. 6541–45; TsGAOR, f. 3348, op. 1, ed. khr. 3, l. 40–41.

21. TsGAOR, f. 3348, op. 1, d. 1140, 1149.

22. Palchinsky, *Torgovye porty Evropy*, 4 vols. (Khar'kov, 1913).

23. This was one of Palchinsky's favorite phrases, and was probably borrowed from Kropotkin. For another example, see his "Rol' i zadachi inzhenerov v ekonomicheskom stroitel'stve Rossii," TsGAOR, f. 3348, op. 1, ed. khr. 695, l. 4. For an example of Kropotkin's use of similar phrases, see his *Fields, Factories and Workshops* (1900; rept. New York: Gordon Press, 1974), p. x.

24. Thomas P. Hughes, *American Genesis: A Century of Invention and Technological Enthusiasm* (New York: Penguin Books, 1990), p. 3 and *passim*.

25. For examples, see Palchinsky, "Sibirskaia kamennougol'naia promyshlennost' i zheleznodorozhnoe khozaistvo," *Zapiski imperatorskago Rossiiskago tekhnicheskago obshchestva* (1908); "Russkii marganets i ego soperniki," *Gorno-Zavodskii listok* 27 (1908); "Vozmozhnost' eksporta Donetskago uglia vo Frantsiiu cherez Mariupol'-Marsel'," *Gorno-Zavodskii listok* 84 (1908); "Eksport zagranitsu produktov gornoi i gornozavodskoi promyshlennosti iuga Rossii," *Izdatel'stvo soveta s"ezda gornopromyshlennosti iuga Rossii* (1911–1913); "Russkii antratsit na turetskom rynke," *Gornozavodskoe delo* 13 (1912); TsGAOR, f. 3348, op. 1, ed. khr. 3, l. 40–41.

26. Palchinsky, "Mestorozhdeniia iskopaemago uglia vdol' Sibirskoi zh-d. magistrali i ikh znachenie dlia kraia," *Gornyi zhurnal* 4 (1907), p. 66.

27. Palchinsky, "Zamechaniia po povodu prichin maloi podgotovlennosti k samostoiatel'noi rabote, davaemoi spetsial'nymi vysshimi shkolami molodym inzheneram, i o sposobakh izmeneniia takogo polozheniia" (Khar'kov, 1907), TsGAOR, f. 3348, op. 1, ed. khr. 1, l. 40ff.

28. TsGAOR, f. 3348, op. 1, ed. khr. 3, l. 38. See also Palchinsky, "Russkaia promyshlennost' na mezhdunarodnoi vystavke v Turine v 1911g.," *Gornyi zhurnal* 3 (1911), pp. 290–303.

29. See his book *Sud i raskol'niki-sektanty* (St. Petersburg, 1901).

30. TsGAOR, f. 3348, op. 1, d. 1176, l. 1.

31. TsGAOR, f. 3348, op. 1, d. 1011, l. 26–27.

32. TsGAOR, f. 3348, op. 1, d. 1011, l. 230–231.

33. TsGAOR, f. 3348, op. 1, d. 1011, l. 371–372.

34. He was a member of the board of the Lyssva Mining District Co., Limited, usually called the "Shouvaloff Company." M. J. Larsons, *An Expert in the Service of the Soviet* (London: Ernest Benn Limited, 1929), pp. 199–207. Also TsGAOR, f. 3348, op. 1, ed. khr. 3, l. 38.

35. Lewis H. Siegelbaum, *The Politics of Industrial Mobilization: A Study of the War-Industries Committees* (London: Macmillan, 1983).

36. Robert P. Browder and Alexander F. Kerensky, eds., *The Russian Provisional Government, 1917: Documents* (Stanford: Stanford University Press, 1961), vols. 1–3, pp. 730–731, 764–765, 1270, 1586, 1788–1790.

37. See Alexander Rabinowitch, *The Bolsheviks Come to Power: The Revolution of 1917 in Petrograd* (New York: W. W. Norton, 1976), pp. 280–301.

38. Palchinsky's role at the Winter Palace defense has been mentioned in

the secondary literature perhaps more than any other of his activities. See, for example, Tsuyoshi Hasegawa, *The February Revolution, Petrograd, 1917* (Seattle: University of Washington Press, 1981), p. 335; Alexander Kerensky, *Russia and History's Turning Point* (New York: Duell, Sloan and Pearce, 1965), p. 266; Anton Antonov-Ovseyenko, *The Time of Stalin: Portrait of Tyranny* (New York: Harper and Row, 1980), p. 119; and Richard Pipes, *The Russian Revolution* (New York: Knopf, 1990), p. 489.

39. P. I. Palchinsky, "Poslednie chasy vremennogo pravitel'stva v 1917 godu," *Krasnyi arkhiv* 1:56 (1933), pp. 136–138. This account is based on his notes taken during and immediately after the takeover of the Winter Palace. TsGAOR, f. 3348, op. 1, d. 184, l. 1–2.

40. Ibid.

2. FROM POLITICAL PRISONER TO SOVIET CONSULTANT

1. Alexander Rabinowitch, *The Bolsheviks Come to Power: The Revolution of 1917 in Petrograd* (New York: Norton, 1976), p. 300.

2. Ibid.

3. Ibid., pp. 300–301.

4. Pitirim Sorokin, *Leaves from a Russian Diary* (New York: Dutton, 1924).

5. TsGAOR, f. 3348, op. 1, ed. khr. 38, 17–37 and 830.

6. TsGAOR, f. 3348, op. 1, ed. khr. 830, l. 3–9.

7. Ibid.

8. V. I. Lenin, *Collected Works*, vol. 44 (Moscow: Progress Publishers, 1970), p. 168.

9. TsGAOR, f. 3348, op. 1, d. 1011, l. 504.

10. TsGAOR, f. 3348, op. 1, d. 1011, l. 527.

11. Examples of Lenin's past suspicion of Palchinsky can be found in short references in Lenin, *Collected Works*, vol. 25, pp. 137, 138, 142, 234, 349, 350, 370, 393.

12. TsGAOR, f. 3348, op. 1, d. 1011, l. 522–523.

13. Electrification seized the imagination of many socialists, both in Russia and abroad. Kropotkin wrote in 1899 that electric power would allow the dispersal of the culture of cities, fulfilling the Marxist hope of eliminating the difference between urban and rural areas. See Thomas P. Hughes, "A Technological Frontier: The Railway," in Bruce Mazlish, ed., *The Railroad and the Space Program* (Cambridge, Mass.: MIT Press, 1965), p. 66. The German socialist Karl Ballod wrote about how elec-

trification and socialism were particularly compatible, and Ballod was later cited as one of the inspirations to the alleged Industrial Party in Russia. Atlanticus (pseud. of Karl Ballod), *Gosudarstvo budushchago*, trans. from German, preface by Karl Kautsky (Moscow, 1906). See Jonathan Coopersmith, *The Electrification* of Russia, 1880–1926 (Ithaca, N.Y.: Cornell University Press, 1992), esp. pp. 139–140.

14. "Perechen' uchrezhdenii i glavneishikh voprosov v koikh prinimal uchastie prof. P. A. Pal'chinskii s I/XI–1917g.," TsGAOR, f. 3348, op. 1, ed. khr. 3, l. 46–49.

15. Letter to the author from Paul Avrich, October 31, 1978; and Samuel A. Oppenheim, "Pal'chinskii, Petr Akimovich," *The Modern Encyclopedia of Russian and Soviet History*, vol. 26 (Gulf Breeze, Fla.: Academic International Press, 1982), pp. 188–189.

16. Aleksandr I. Solzhenitsyn, *The Gulag Archipelago*, vol. 2 (New York: Harper and Row, 1975), p. 314.

17. M. J. Larsons (pseudonym of Maurice Laserson), *An Expert in the Service of the Soviet* (London: Ernest Benn, 1929), p. 205.

18. TsGAOR, f. 3348, op. 1, ed. khr. 793, l. 5.

19. TsGAOR, f. 3348, ed. khr. 1, l. 41, 3.

20. TsGAOR, f. 3348, op. 1, ed. khr. 6, l. 1.

21. TsGAOR, f. 3348, op. 1, ed. khr. 901, l. 1–4.

22. Loren R. Graham, *The Soviet Academy of Sciences and the Communist Party, 1927–1932* (Princeton, N.J.: Princeton University Press, 1967), pp. 100–111, 135–137.

23. TsGAOR, f. 3348, op. 1, ed. khr. 589.

24. TsGAOR, f. 3348, op. 1, ed. khr. 553, l. 11–17.

25. Palchinsky, *Poverkhnost' i nedra* 2:18 (1926), p. 1. Palchinsky gave an economic analysis of roofing materials, showing that in different places at different times the optimal material might be any one of metal, tile, brick, glass, wood, cement, clay, or stone. Palchinsky, *Osnovnye zadachi razvitiia promyshlennosti stroitel'nykh materialov* (Leningrad, 1924).

26. TsGAOR, f. 3348, op. 1, ed. khr. 563, l. 1.

27. TsGAOR, f. 3348, op. 1, ed. khr. 563, l. 25.

28. TsGAOR, f. 3348, op. 1, ed. khr. 566, l. 48.

29. Ibid.; ed. khr. 558, l. 32, 42.

30. See the journal *Poverkhnost' i nedra*. See also "Materialy redaktsii ezhemesiachnogo nauchnogo tekhno-ekonomicheskogo zhurnala *Poverkhnost' i nedra*, redaktorom kotorogo byl Pal'chinskim, P. A. (1916–1928),"

TsGAOR, f. 3348, op. 1, ed. khr. 187–246, and "Materialy otnosiash-chiesia k periodu deiatel'nosti Pal'chinskogo P. A. v institute izuchenii *Poverkhnost' i nedra* (1916–1926)," TsGAOR, f. 3348, op. 1, ed. khr. 247–517.

31. Kendall E. Bailes, "The Politics of Technology: Stalin and Technocratic Thinking among Soviet Engineers," *The American Historical Review* 79 (1974), p. 452, citing *Vestnik inzhenerov* 1–2 (1924), pp. 9–11. Bailes's report that Palchinsky translated a book by Herbert Hoover must be mistaken, as no record of that book can be found in the Lenin Library in Moscow nor in Palchinsky's very careful listings of his writings and translations.

32. Palchinsky, "Rol' i zadachi inzhenerov v ekonomicheskom stroitel'stve Rossii," TsGAOR, f. 3348, op. 1, ed. khr. 695.

33. In 1926 Palchinsky got into a controversy with the Council on the Petroleum Industry over whether natural or artificial asphalt was best. Palchinsky opposed the council's blanket order to use only artificial asphalt, saying that the decision depended on local conditions. He evidently lost this debate. TsGAOR, f. 3348, op. 1, ed. khr. 552, l. 78–91.

34. TsGAOR, f. 3348, op. 1, ed. khr. 717.

35. "Otnositel'noe znachenie krupnykh, srednikh i melkikh predpriiatii v kamennougol'noi promyshlennosti Velikobritanii" (Khar'kov, 1911).

36. TsGAOR, 3348, op. 1, ed. khr. 525, l. 264.

37. Central Government Historical Archive (TsGIA), f. 90, op. 1, d. 145, l. 47–49.

38. Palchinsky, "Gornaia ekonomika," *Poverkhnost' i nedra* 2:18 (1926), p. 12.

39. TsGAOR, f. 3348, op. 1, ed. khr. 751, l. 2.

40. Ibid.

41. Palchinsky, "Gornaia ekonomika," pp. 14–15.

42. TsGAOR, f. 3348, op. 1, ed. khr. 760.

43. Palchinsky, "Gornaia ekonomika," pp. 14–15.

44. Ibid., p. 17.

45. TsGAOR, f. 3348, op. 1, ed. khr. 525, l. 190–191.

46. Kendall Bailes, "Aleksei Gastev and the Soviet Controversy over Taylorism," *Soviet Studies* 3 (1977), pp. 373–394; Zenovia Sochor, "Soviet Taylorism Revisited," *Soviet Studies* 2 (1981), pp. 246–264.

47. V. I. Lenin, "Variant stat'i 'Ocherednye zadachi sovetskoi vlasti'," *Polnoe sobranie sochinenii*, 5th ed. (Moscow: Izdatel'stvo politicheskoi literatury, 1969), p. 141.

48. René Fulop-Miller, *Geist und Gesicht des Bolschewismus: Darstellung und Kritik des kulturellen Lebens in Sowjet-Russland* (Vienna: Amalthea-Verlag, 1926), p. 29.

49. TsGAOR, f. 3348, op. 1, ed. khr. 693, l. 9–10.

50. TsGAOR, f. 3348, op. 1, d. 562, l. 1.

51. TsGAOR, f. 3348, op. 1, ed. khr. 41, l. 52–53.

52. TsGAOR, f. 3348, op. 1, ed. khr. 46, l. 31.

53. Palchinsky, "Mestorozhdeniia iskopaemago uglia vdol' zh.-d. magistrali i ikh znacheniia dlia kraia," *Gornyi zhurnal* 4 (1907), pp. 66–70.

54. TsGAOR, f. 3348, op. 1, ed. khr. 550, l. 44.

55. Palchinsky, "Ekonomicheskaia geologiia," *Poverkhnost' i nedra* 4 (1926), p. 5, and TsGAOR, f. 3348, op. 1, ed. khr. 695.

56. Edwin T. Layton, Jr., *The Revolt of the Engineers: Social Responsibility and the American Engineering Profession* (Cleveland and London: Case Western Reserve University Press, 1971), p. 6.

57. Editorial, *Poverkhnost' i nedra* 1 (1926), p. 6.

58. Palchinsky, "Otnositel'noe znachenie krupnykh, srednikh i mel'kikh predpriiatii v kamennougol'noi promyshlennosti Velikobritanii" (Khar'kov, 1911), pp. 1–9.

59. For example, John Scott, *Behind the Urals: An American Worker in Russia's City of Steel* (Bloomington: Indiana University Press, 1989), and Michael Gelb, ed., *An American Engineer in Stalin's Russia: The Memoirs of Zara Witkin, 1932–1934* (Berkeley: University of California Press, 1991).

60. Quoted in Kendall Bailes, *Technology and Society under Lenin and Stalin: Origins of the Soviet Technical Intelligentsia, 1917–1941* (Princeton,, N.J.: Princeton University Press, 1978), pp. 88 and 148.

61. Ibid., p. 160.

62. Conversation with Daniel Aleksandrov, Leningrad, October 1990, and Moscow, October 1991.

63. Bailes, *Technology and Society*, p. 466.

64. TsGAOR, f. 3348, op. 1, ed. khr. 57, l. 105–109.

65. Bailes, *Technology and Society*.

66. Bailes, "The Politics of Technology," pp. 453–454.

67. Ibid., p. 455.

68. Ibid., pp. 456–458.

69. Ibid., p. 458.

70. The best biography of Bukharin is Stephen F. Cohen, *Bukharin and the Bolshevik Revolution* (New York: Knopf, 1973).

71. Bailes, "The Politics of Technology," p. 463.

72. The best source on the Shakhty Trial is Bailes, *Technology and Society*, pp. 69–94.

73. *Protsess "Prompartii" 25 noiabria–7 dekabria 1930 g. Stenogramma sudebnogo protsessa i materialy priobshchennye k delu* (Moscow, 1931).

74. Aleksandr Solzhenitsyn, *The First Circle*, trans. Thomas P. Whitney (New York: Harper and Row, 1968).

75. Solzhenitsyn, *The Gulag Archipelago* (New York: Harper and Row, 1974), vol. 2, p. 637, quoting from the manuscript section of the Lenin Library, Collection 410, card file 5, storage unit 24.

76. Solzhenitsyn, *The Gulag Archipelago*, vol. 1, p. 74.

77. *Materialy k otchetu TsK VKP(b) XVI s"ezdu VKP(b) sostavlennyi OGPU*, INION AN SSSR.

78. Palchinsky, "Gornaia ekonomika," *Poverkhnost' i nedra* 2:18 (1926), p. 12.

79. TsGAOR, f. 3348, op. 1, ed. khr. 297, l. 18.

80. "Klub gornykh deiatelei (KGD) v Moskve," *Poverkhnost' i nedra* 2:18 (1926), p. 35.

3. EARLY SOVIET INDUSTRIALIZATION

1. For an entertaining account of Hugh Cooper's participation, see Thomas P. Hughes, *American Genesis: A Century of Invention and Technological Enthusiasm* (New York: Penguin Books, 1989), pp. 264–269.

2. Anne D. Rassweiler, *The Generation of Power: The History of Dneprostroi* (New York: Oxford University Press, 1988), p. 56.

3. The completed hydroelectric station used nine 85,000-horsepower turbines built by the Newport News Shipbuilding and Drydock Company.

4. Rassweiler, *The Generation of Power*, pp. 45–47.

5. Boris Komarov, *The Destruction of Nature in the Soviet Union* (White Plains, N.Y.: M. E. Sharpe), p. 57.

6. Ibid.

7. Rassweiler, *The Generation of Power*, pp. 120–122.

8. Komarov, *The Destruction of Nature*, p. 57.

9. Palchinsky, "Gornaia ekonomika," *Poverkhnost' i nedra* 1:29 (1927), p. 9.

10. Michael Gelb, ed., *An American Engineer in Stalin's Russia: The Memoirs of Zara Witkin, 1932–1934* (Berkeley: University of California Press, 1991), pp. 232–245.

11. Quoted in Gelb, *An American Engineer in Stalin's Russia*, p. 234.

12. Stephen Kotkin, *Steeltown, USSR: Soviet Society in the Gorbachev Era* (Berkeley: University of California Press, 1991), p. 208.

13. Ibid., p. 209.

14. Ibid., p. 121.

15. Ibid., pp. 227–228.

16. Ibid., p. 228.

17. John Scott, *Behind the Urals: An American Worker in Russia's City of Steel*, 2nd enlarged ed., prepared by Stephen Kotkin (Bloomington: Indiana University Press, 1989).

18. Ibid., p. xxii, as quoted from John Scott's original manuscript.

19. Kotkin, *Steeltown, USSR*, p. 254.

20. Amabel Williams-Ellis, "Introduction," *Belomor: An Account of the Construction of the New Canal between the White Sea and the Baltic Sea* (New York: Harrison Smith and Robert Haas, 1935), p. vi.

21. Boris Souvarine, *Stalin: A Critical Survey of Bolshevism* (New York: Longmans, Green, 1939), p. 504.

22. Aleksandr Bogdanov, *Red Star: The First Bolshevik Utopia*, trans. Charles Rougle, ed. Loren R. Graham and Richard Stites (Bloomington: Indiana University Press, 1984). This volume contains both *Red Star* and *Engineer Menni*.

23. TsGAOR, f. 3348, op. 1, ed. khr. 695, l. 19.

24. Aleksandr I. Solzhenitsyn, *The Gulag Archipelago*, vol. 2 (New York: Harper and Row, 1975).

25. M. Gor'kii, L. L. Averbakh, and S. G. Firin, eds., *Belomorsko-Baltiiskii Kanal imeni Stalina: Istoriia stroitel'stva* (Moscow, 1934), p. 75.

26. Quoted in Solzhenitsyn, *The Gulag Archipelago*, vol. 2, p. 99.

27. Ibid., pp. 100–102.

4. TECHNOCRACY, SOVIET STYLE

1. William A. Wood, *Our Ally, The People of Russia* (New York: Scribner's, 1950), pp. 127–128.

2. Kendall Bailes, *Technology and Society under Lenin and Stalin: Origins of the Soviet Technical Intelligentsia, 1917–1941* (Princeton, N.J.: Princeton University Press, 1976).

3. Sheila Fitzpatrick, *The Commissariat of the Enlightenment* (Cambridge, England: Cambridge University Press, 1970).

4. Nicholas DeWitt, *Education and Professional Employment in the U.S.S.R.* (Washington, D.C.: National Science Foundation, 1961), pp. 209, 225.

5. Ibid., p. 217.

6. Ibid., p. 216.

7. Ibid., p. 225.

8. Harley Balzer, "Engineers: The Rise and Decline of a Social Myth," in Loren Graham, ed., *Science and the Soviet Social Order* (Cambridge, Mass.: Harvard University Press, 1990), p. 152.

9. DeWitt, *Education and Professional Employment*, p. 226.

10. *Administration of Teaching in Social Sciences in the U.S.S.R. (Syllabi for Three Required Courses: Dialectical and Historical Materialism, Political Economy, and History of the C.P.S.U., Moscow, 1957)* (Ann Arbor: University of Michigan, 1960).

11. Kendall Bailes, "The Politics of Technology: Stalin and Technocratic Thinking among Soviet Engineers", *American Historical Review* 79 (1974), p. 469.

12. I am grateful to Thomas P. M. Barnett for sharing his research paper, "Post-Stalinist Trends in the Soviet Politburo: The Development of Technocracy?" (Harvard University Government Department, January 29, 1987).

13. Quoted in Bailes, *Technology and Society*, p. 419.

14. *New York Times*, March 31, 1992, p. A7.

15. An article on Soviet nuclear power that takes a particularly historical view is Paul Josephson, "The Historical Roots of the Chernobyl' Disaster," *Soviet Union/Union Sovietique* 13:3 (1986), pp. 275–299.

16. "Debating the Need for River Diversion," *Current Digest of the Soviet Press* (March 19, 1986), p. 1. An excellent analysis of the controversy over river diversion is Robert G. Darst, Jr., "Environmentalism in the USSR: The Opposition to the River Diversion Projects," *Soviet Economy* (July–September 1988), pp. 223–252.

17. See "The Virgin Land Debate," in Werner G. Hahn, *The Politics of Soviet Agriculture, 1960–1970* (Baltimore and London: Johns Hopkins University Press, 1972), pp. 26–33.

18. See Mikhail Geller and Alexander Nekrich, *Utopiia u vlasti: Istoriia Sovetskogo Soiuza c 1917 goda do nashikh dnei* (London: Overseas Publications Interchange, 1989), p. 34, and I. V. Stalin, *Sochineniia*, vol. 2 (Stanford, Calif.: The Hoover Institution, 1967), p. 206. I am grateful to Alexander Nekrich for bringing this incident to my attention.

19. Harley Balzer, "Engineers: The Rise and Decline of a Social Myth," in Graham, *Science and the Soviet Social Order*, pp. 141–147.

20. Loren R. Graham, "Reorganization of the USSR Academy of Sciences," in Peter Juviler and Henry Morton, eds., *Soviet Policy-Making* (New York: Praeger, 1967), pp. 133–163.

21. "Appeal of Soviet Scientists to the Party-Government Leaders of the U.S.S.R.," *Survey* 76 (1970), pp. 160–170.

22. Josephson, "Historical Roots of the Chernobyl' Disaster," pp. 275–299.

5. CONTEMPORARY ENGINEERING FAILURES

1. *The Great Baikal-Amur Railway* (Moscow: Progress Publishers, 1977), p. 8.

2. V. Perevedentsev, "Where Does the Road Lead?" *Current Digest of the Soviet Press* 40:46 (1988), from *Sovetskaia kul'tura*, October 11, p. 3.

3. *The Great Baikal-Amur Railway*, p. 1.

4. Ibid., pp. 85-86. A *kabarga* is a musk deer native to Siberia.

5. *CDSP* 39:34 (1987), from *Izvestiia*, August 21, p. 2.

6. *CDSP* 39:23 (1987), from *Pravda*, June 11.

7. *CDSP* 41:17 (1989), from *Pravda*, April 26, p. 3.

8. Ibid.

9. Ibid.

10. *CDSP* 39:10 (1987), from *Sotsialisticheskaia industriia*, February 11, p. 2.

11. V. Khatuntsev, "Why the Young Main Line Is Not Operating at Full Capacity," *Pravda*, June 11, *CDSP* 39:23 (1987), p. 21.

12. Boris Komarov, *The Destruction of Nature in the Soviet Union* (White Plains, N.Y.: M. E. Sharpe, 1980), pp. 116–127.

13. Conversations with Vladimir Sangi, President of the Peoples of the North, Moscow, December 1990 and October 1991.

14. Perevedentsev, "Where Does the Road Lead?", p. 3.

15. A notable exception is Paul R. Josephson, "The Historical Roots of the Chernobyl Disaster," *Soviet Union/Union Sovietique* 13:3 (1986), pp. 275–299.

16. David R. Marples, *Chernobyl and Nuclear Power in the USSR* (New York: St. Martin's, 1986), p. 74.

17. David R. Marples, *The Social Impact of the Chernobyl Disaster* (New York: St. Martin's, 1988), p. 3.

18. *The New York Times*, March 25, 1992, p. A7, and November 8, 1992, pp. A1, A14.

19. Josephson, "Historical Roots of the Chernobyl Disaster," p. 278.

20. Ibid., p. 283.

21. N. Dollezhal and Iu. Koriakin, *Kommunist* 14 (September 1979).

22. Grigori Medvedev, *The Truth about Chernobyl* (New York: Basic Books, 1989), p. 2.

23. *USA Today*, March 25, 1992, p. 9A.

24. His closest advisor was Aleksandr Yakovlev, a historian who has been called "the architect of *perestroika.*"

25. David R. Marples, *Ukraine under Perestroika: Ecology, Economics, and the Workers' Revolt* (New York: St. Martin's, 1991), p. 188.

26. N. V. Melnikov, O. D. Didin, and A. T. Ayruni, "Results of Research on the Problem of Sudden Methane and Coal Outbursts in the U.S.S.R.," *Systems for Stimulating the Development of Fundamental Research*, report of the U.S.-U.S.S.R. Working Subgroup on Systems for Stimulating the Development of Fundamental Research of the National Academy of Sciences/National Research Council, Commission on International Relations (Washington, D.C.: National Academy of Sciences, 1978), pp. X-1–X-39. I was a member of this working group.

27. *MESA Safety Review*, 1974, 1975 (Washington, D.C.: U.S. Department of the Interior Mining Enforcement and Safety Administration).

28. *Injury Experience in Coal Mining, 1980* (Washington, D.C.: Mine Safety and Health Adminstration, 1981).

29. Marples, *Ukraine under Perestroika*, p. 197.

30. Ibid.

31. Ibid., pp. 197–198.

32. Ibid., p. 209.

33. Lewis H. Siegelbaum, "Behind the Soviet Miners' Strike," *The Nation*, October 23, 1989, p. 452.

34. Marples, *Ukraine under Perestroika*, pp. 208, 210.

35. Ibid., p. 201.

36. Palchinsky, "Gornaia ekonomika," *Poverkhnost' i nedra* 2 (1926), p. 14.

EPILOGUE

1. Vaclav Havel, "The End of the Modern Era," *The New York Times*, March 1, 1992, p. 15.

2. The statistics were cited by the Soviet Minister of Health, E. I. Chazov,

"Speech," *Current Digest of the Soviet Press* 40:27 (1988), from *Pravda*, June 30, pp. 4, 9.

3. Edwin T. Layton, Jr., *The Revolt of the Engineers: Social Responsibility and the American Engineering Profession* (Cleveland and London: Case Western Reserve, 1971).

4. I am indebted to Victoria Post Ranney and George A. Ranney, Jr., for insights on the qualities of mining engineers.

5. Solzhenitsyn draws this conclusion in *The Gulag Archipelago*, vol. 1 (New York: Harper and Row, 1975), p. 375.

Acknowledgments

The person to whom I owe the greatest debt in encouraging me to undertake this book is Howard Boyer, former Science Editor at Harvard University Press. Howard listened to my ideas on the subject, read the earliest versions of the manuscript, and greatly improved the final product. Aida Donald, Jennifer Snodgrass, and the entire Harvard University Press staff have helped transform the idea of this book into a reality, and I am most grateful to them for continuing support and encouragement.

The institution to which I owe the greatest debt in writing this book is the John D. and Catherine T. MacArthur Foundation. I have been fortunate to receive a grant from the MacArthur Foundation to pursue my interests in the history of Russian science and technology and to conduct research in Moscow. With its support I have organized a series of workshops in Moscow and Cambridge, Massachusetts, entitled "Science and Technology with a Human Face," at which scholars have explored social issues connected to science and technology. Many of these issues are the same ones in which Palchinsky, many years ago, was interested.

In the Moscow archives I have worked with a research assistant, Vyacheslav Gerovitch, who brought to the project both sophistication and industry. Our discussions of Palchinsky and his vicissitudes in the hallways of the archives, on the streets of Moscow, and in Cambridge have been a personal delight and a professional boost.

I am grateful to Boris Kozlov, director of the Institute of the History of Science and Technology of the Russian Academy of Sciences, for assistance to me and my graduate students, and for obtaining several photographs for the book. In Cambridge, Gregory Crowe, Charles Holtzman, and Jennifer Haywood have helped me in many ways, from working in the libraries to arranging research trips to Moscow. The director of the Program on Science, Technology and Society at MIT, Merritt Roe Smith, has helped me to understand more fully the subtleties of the history of technology. A similar role has been played by Thomas P. Hughes, Professor of the History of Technology at the University of Pennsylvania, whose own work has inspired me. At Harvard University, where I have taught part-time in recent years, the History of Science Department and the Russian Research Center have provided significant institutional support.

My wife, Patricia Albjerg Graham, has always been a fellow scholar on whom I have tried most of my ideas, both the good ones and the bad ones. She has helped me to see the difference, a service for which no compensation is sufficient.

Index

Dnieper River hydroelectric plant (Dneprostroi) (*continued*) 81, 83, 99; working conditions, 52, 54–55, 100
Don Basin coal industry (Ukraine), 8–10, 13, 15, 46, 48, 81, 104; workers' strike, 93–97
Dubovik, A. S., 94
Dudintsev, Vladimir, 78

Engelmeier, P. K., 43–44
Engels, Friedrich, 72
Engineering: education, 14–15, 42, 68–71, 72–73, 74, 77, 99, 105; "humanitarian," 39–41, 102; role of, 40–41, 43, 62, 67–68, 103; Communist Party and, 67–68, 73, 78, 79, 105; dissidence in, 77–79
Engineer Menni (Bogdanov), 62
Engineers' Herald (*Vestnik inzhenerov*), 43
Environmental issues, 55, 88–89, 101

First Circle, The (Solzhenitsyn), 45
Five-Year Plans, 3, 49, 50, 83, 84, 90, 100; First Five-Year Plan, 33, 49, 51, 61, 81
Ford, Henry, 38–39, 102
France, 103

Gary, Ind., steel industry, 56, 57, 78, 103
Gastev, Aleksei, 39
Germany, 21, 58, 77, 103; Nazi, 100
Gerschenkron, Alexander, 3
Glasnost', 2, 93
Gorbachev, Mikhail, 3, 49, 75, 92–93; reforms, 2, 58, 87, 96–97
Gorky atomic power station, 86
Gorlovka Mine, 9–10
Gosplan (State Planning Commission), 30, 32, 35
Great Depression, 50, 60
Gubkin, I. M., 27, 32
Gulag Archipelago (Solzhenitsyn), 1

Havel, Vaclav, 99, 100, 102, 103
Hitler, Adolf, 4, 58
Hoover, Herbert, 103

Hughes, Thomas, 14
"Humanitarian engineering," 39–41, 102
Hungary, 84
Hydroelectric power projects, 54, 62. *See also* Dnieper River hydroelectric plant

Ignalina atomic power station, 86
Industrial Party, 43, 46
Industrial Party Trial, 45, 47, 67, 106
INION library, 46
Institute of Labor, 39
Institute of the Surface and Depths of the Earth, 20, 32, 34, 38, 46, 47
Irkutsk, Siberia, 12, 13, 15
Izvestiia, 1

Jean Christophe (Rolland), 29
Josephson, Paul, 91

Kalinnikov, I. A., 43
Karelia region, 61–62
Kazan, Russia, 6, 7
Kerensky, Aleksandr, 23, 26
Khrushchev, Nikita, 69, 72, 76, 84, 96, 105
Khrustalev, N. I., 62–63
Kiev, 20
Kishkin, Nikolai, 21, 25
Klasson, R. E., 52–53
Kokoshkin, Fedor, 24, 26
Komintern, 43
Kommunist (journal), 91
Komsomol (Young Communist League), 83
Kotkin, Stephen, 60–61
Kovalevskii, V. I., 10
Kropotkin, Peter, 10, 11, 17, 21, 24, 29, 45
Krzhizhanovskii, Gleb M., 30, 32
Kumsa River, 63
Kuznetsk (later Novokuznetsk), Siberia, 51, 82

Labor: forced, 59–60, 61, 64, 83, 86–87; volunteer, 59–60, 61, 83; apathy of, 60, 101–102; strike, 93–97

Russian Revolution, 12
Russian Technical Society, 29, 39
Rutenberg, Peter, 21, 25
Rybinsk hydroelectric power plant, 54
Rykov, Aleksei Ivanovich, 43, 44

St. Petersburg, 5, 13, 15, 28, 61, 92
Sakharov, Andrei, 79
Scientific Technical Administration, 44
Scott, John, 60–61
Secret police (Cheka), 1, 6, 26, 28, 46, 105
Seeds, Nellie, 60
Segezha River, 63
Semenov, N. N., 78
Shell Oil, 46
Shenk family, 18, 19, 20
Shingarev, Andrei, 24, 26
Siberia, 10, 56, 84, 88, 93, 101; Palchinsky exiled in, 12, 13, 20, 46
Socialist Revolutionary Party, 12, 21, 40
Solzhenitsyn, Aleksandr, 1, 45, 65
Sorokin, Pitirim, 24, 25
Southern Yakutia, 89
Stalin, Joseph, 1, 41–44, 50, 58, 69, 77, 87, 105; industry and, 52, 59, 62, 65, 83, 99–100
State Bank, 30
Styrikovich, M. A., 92
Sukhomlinov, Vladimir, 24
Supreme Council of the National Economy (VSNKh), 33, 44
Surface and Depths of the Earth, The (journal), 20

Taylor, Frederick, 38–39, 102
Tereshchenko, Mikhail, 24, 25
Toepfer, Klaus, 92
Tolstoy, Leo, 16–17
Tomsky, M. P., 96
Trans-Siberian Railway, 82, 89
Trotsky, Leon, 40, 46, 73
Tsarist regime, 6, 8, 12, 14, 20, 68, 100, 104

Udokan copper deposits, 82, 89
Ukraine, 13, 56. *See also* Don Basin coal industry
United States, 38–39, 41, 56, 69, 94, 103
United States Steel, 57, 78
Upper Angara River, 82

Verzhbitskii, K. A., 63
Viazemskii, O. V., 62–63
Vitkovskii, D. P., 64
Volgadonsk, 91
Volga River, 86
Vyg River, 63

War Industry Committee, 20
Wells, H. G., 43
White Sea, 50, 61
White Sea Canal (Belomorstroi), 50–51, 61–65, 81, 85, 99
Winter Palace, 21–22, 23

Yagoda, Genrikh, 65
Yeltsin, Boris, 3, 49

Zhiguli automobiles, 83, 88–89, 101
Zinoviev, Grigorii, 26
Zubrik, K. M., 63